The Joy
of
Sweat

The Joy *of* Sweat

....

THE STRANGE SCIENCE OF PERSPIRATION

....

Sarah Everts

W. W. NORTON & COMPANY
Independent Publishers Since 1923

For information about permission to reproduce selections from this book, write to
Permissions, W. W. Norton & Company, Inc., 500 Fifth Avenue, New York, NY 10110

For information about special discounts for bulk purchases, please contact
W. W. Norton Special Sales at specialsales@wwnorton.com or 800-233-4830

Manufacturing by Lake Book Manufacturing
Book design by Chris Welch
Production manager: Anna Oler

Library of Congress Cataloging-in-Publication Data

Names: Everts, Sarah, author.
Title: The joy of sweat : the strange science of perspiration /
Sarah Everts.
Description: First edition. | New York, NY : W. W. Norton & Company, [2021]
| Includes bibliographical references and index.
Identifiers: LCCN 2020053349 | ISBN 9780393635676 (hardcover) |
ISBN 9780393635683 (epub)
Subjects: LCSH: Perspiration.
Classification: LCC QP221 .E94 2021 | DDC 612.7/93—dc23
LC record available at https://lccn.loc.gov/2020053349

W. W. Norton & Company, Inc., 500 Fifth Avenue, New York, N.Y. 10110
www.wwnorton.com

W. W. Norton & Company Ltd., 15 Carlisle Street, London W1D 3BS

1 2 3 4 5 6 7 8 9 0

For Quinn

CONTENTS

The Joy of Sweat

In the summer of 1996, a woman walked into the dermatology office at Tygerberg Hospital in suburban Cape Town with an unusual complaint: Her sweat was red.

She was understandably alarmed. But the medical team? They were intrigued.

"It was fantastically interesting," says Corena de Beer, a scientist who analyzed the case. "We spent months trying to figure out what was happening. Here was a healthy person, a nurse, in her twenties. The moment she started sweating, pink spots would appear on her white uniform." By the end of a shift, the nurse's underwear and uniform were sometimes bright red, particularly around the collar, back, and armpits. "Every night she soaked her stained clothes for 2 to 3 hours before washing them just to get the color out," de Beer says. "Not only was it very disconcerting to have something so abnormal happening to her body, but the nurse was also concerned about her job. In a hospital, nurse uniforms need to be white. She felt the red sweat was socially and professionally unacceptable."

Dermatologists see all manner of skin curiosities, but red sweat is so unusual that de Beer and dermatologist Jacques Cilliers published a scientific paper about it: "The Case of the Red Lingerie— Chromhidrosis Revisited." *Chromhidrosis* is the medical term for colored (*chrom*) sweat (*hidrosis*). This nurse, it turns out, was not the first person to produce pigmented perspiration, nor would she be the last. The medical literature features many reports of sweat that has turned green, blue, yellow, brown, or red, with causes as varied as rare genetic conditions and workplace chemical exposure.

There are even subcategories of the colored sweat phenomenon, including a faux version called *pseudo chromhidrosis*. That's when sweat coming out of a person's pores only acquires a color after it hits the skin surface, a situation that happens occasionally to copper workers. The salt in their sweat oxidizes traces of copper on their skin to create a beautiful, if alarming, green sheen on their bodies like the bluish-green patina on the copper domes of buildings after years of weathering.

But the nurse's perspiration was colored when it emerged from her pores, making it a *true chromhidrosis*. Something red inside her body was making an exit through the skin and its sweat pores. The medical team conducted a thorough examination of the nurse and found she was strikingly healthy, red sweat notwithstanding. Everyone was stumped, de Beer says. For a long time, the medical team had only one serious clue about the cause of her condition: While the nurse had been on a 4-week holiday, her red sweat had abated, nearly disappearing. But when she got back to her work routine, the red sweat returned. "We started wondering if it might be stress-related. So we checked everything—her liver, her endocrine system," de Beer adds. "But they were all totally fine."

In the end, the medical team discovered the source of her red sweat thanks to an accidental observation. At a follow-up appointment, the nurse showed up at the dermatology practice with brownish-red stains on her fingers, similar to the nicotine stains from a cigarette. "But we knew she wasn't a smoker," de Beer says. "And that's when the penny dropped." The stains on her fingers came from a pre-appointment snack: a South African corn chip called NikNaks Spicy Tomato.

It turns out the nurse had an obsessive predilection for the spicy tomato treat, which her medical team described in their scientific article as a "six-month fetish." Calling her love of the corn chips a fetish was not hyperbolic. The nurse told the doctors she was "indulging in 500 to 2,500 grams per week over an extended period." That's between 1 and 5½ pounds of chips a week. Given that a bag of NikNaks weighs 55 grams, the nurse had been eating up to 45 bags a week, or about 6 bags a day.

"All South Africans eat NikNaks," de Beer says. "They are salty and tasty. My favorite is the cheesy flavor. We just don't all eat the same volumes as her."

The medical team suspected that something red in the NikNaks was coming out in the nurse's sweat—maybe the paprika or tomato flavoring or perhaps a red dye. The team contacted Simba, the company that produces NikNaks, to get a list of ingredients and found that the red pigmentation in the nurse's sweat was the same as the color of the tomato ingredients added to the corn snack. "At first the nurse couldn't believe what we were telling her," de Beer says. "She was skeptical that [Spicy Tomato] NikNaks were the problem. But we put her on an elimination diet, and in a few months the red sweat was completely gone." Soon the nurse was back to having only the usual complaints about sweat on a hot day: sticky clothing, inconvenient wet patches, a dank aroma.

. . . .

Until I read about the South African nurse, I had the naive and erroneous impression—even with a master's degree in chemistry and more than a decade of experience as a science journalist—that sweat was just a banal mix of salt and water. Plus stink molecules, of course, and possibly pheromones. But to think that our bodies leak secrets about our vices was both fascinating, and, well, alarming.

If a person's chip fetish could be exposed in her sweat, what other privileged information about biology and behavior might also be pouring out? Might Big Brother monitor the food we eat and the drugs we take by sampling our perspiration? Could doctors assess our health or diagnose disease by taking sweat samples? Were we leaving damning traces of ourselves and our vices every time we touched something with sweaty hands? Were these secrets discoverable by analyzing fingerprints, which are inked with sweat? (Spoiler: Yes they are.) Reading about the case of the red perspiration got me hooked on sweat science, like a night nurse on NikNaks.

I couldn't stop thinking about what secrets I was leaving behind in my perspiration—especially given that I tend to lose a lot of it. I have long harbored a nagging worry that I might sweat more than average. When it is hot out, I seem to be the first in any group to glisten. At the gym, I often reach for a towel before the warm-up is over. During hot yoga, I surreptitiously peer at my neighbors' mats, looking for evidence that others are also dripping on their mats when I should be focusing on my own downward dog. The utter absurdity of this scene—someone obsessing about their own perspiration during an intentionally sweaty activity that is ultimately supposed to leave one calm and grounded—is not lost on me. I decided to turn my preoccupation with perspiration into professional curiosity. I sought out people who sniff armpits for work and others who do so in search of romance. I spoke with scientists who study sweat—from the molecules that make us stink to the ones that might get us arrested.

As I began to dig into sweat research, I learned that some people sweat so much, they have difficulty holding a pencil or a cell phone because the objects slip right out of their hands. Sweating can be so socially and professionally debilitating that some individuals struggle with depression and anxiety. And some turn to invasive surgeries—such as cutting the ganglia associated with the spinal cord—to try curbing perspiration.

As a biological process that affects—some might say afflicts—us all, sweating is curiously dramatic. It's such an oddly flamboyant way to control body temperature: a coordinated breach of the skin by the salty ocean inside, so that cooling floods can sweep us back from the brink of feverish delirium and death during hot weather.

Sweating isn't just splashy and conspicuous; it's also so very *human*. Most other animals do not regulate their body temperature by sweating. In fact, some evolutionary biologists argue that our ability to perspire has helped humans to dominate the natural world. Of course, that's cold comfort when your attempt at a crisp professional look is undermined by raunchy wet patches. Or when you can't ever wear a well-fitted suit because it will be drenched before you reach your destination.

Which is why blocking sweating and its smelly consequences is such big business. The world spends $75 billion a year on antiperspirants and deodorants in an attempt to pretend that we do not sweat and that this nonexistent sweat has no odor. No one is duped. But in Western society, we've made hiding our biological nature a matter of social acceptance: This vital life process, one that we all possess, one that helps make us human, is deemed embarrassing and unprofessional. How did that come to be?

Ironically, while we go to great lengths to limit our perspiration, at the same time many of us pay to sweat in vast quantities. There's an unending supply of fitness fads and exercise routines guaranteed to break a sweat; few exercise addicts are satisfied with their workouts if their T-shirts stay dry. Many cultures also have ritualistic sweating ceremonies—if not in modern times, then at some point in their history. There are marbled *hammams* found across the Middle East; Indigenous Peoples throughout the Americas have sweat lodges; Koreans frequent *jjimjilbangs*; Russians drink vodka in *banyas*; and the Finns have exported saunas across the Western world. I began to wonder if there is something in the human psyche that actually craves a good sweat.

....

Why don't we delight in our ability to produce perspiration the way we revel in the ability of a spider to produce silk? I'd wager your average spider isn't self-conscious about the sticky stuff flowing from *her* body.

As the medical historian Michael Stolberg puts it, "Far more than one would expect from a seemingly innocuous, bland, watery fluid, sweat is associated with shame and embarrassment, with pollution and stench, but also with purification, sexual attraction and masculinity." That's a lot of emotional baggage for one bodily fluid to carry.

Wouldn't it be better to find serenity instead of shame in all the sweating that we do? It's not like we're going to evolve an alternative strategy for temperature control any time soon. And it's not like we're going to be sweating less in the future, given the reality of global warming. Our ability to sweat may be foundational to the resilience we'll need to get through the coming climate apocalypse.

Sweat may be sticky, stinky, and gross, but it's also one of our most fascinating and little understood secretions. This book, more than anything else, is a perspiration pep talk. Sweat has received enough side-eye. It's time we all found some joy in sweat.

Part I

The Science of Sweat

1

TO SWEAT IS HUMAN

Being alive is hot business. When you spend a day sitting around doing absolutely nothing, thoroughly enjoying the fact that your most challenging physical activity is lifting chips to your mouth as you binge on Netflix, your body still produces as much heat as a 60-watt light bulb. That's if you are a smallish person. If you are big and burly, you will be as hot as a 100-watt light bulb. Our bodies can't help but exude heat even in a state of utter relaxation because our cells are workaholics, diligently dealing with the never-ending to-do list involved in keeping us alive: breaking down nutrients, shuttling oxygen around, building hormones, copying DNA, fighting pathogens. Taking care of these tasks requires billions of chemical reactions. And many of these reactions produce heat. That internal heat makes us warm.

If you begin moving around—to grab the food delivery from the front door or to venture outside with the dog—the added effort of your muscles dials up heat production. Start to jog down the street and your body's temperature begins to soar. Should you sprint for the bus or after an escaped dog, your internal temperature could easily reach life-threatening levels if unchecked. Dying from heatstroke is a miserable way to go: Multiple organs fail as your cell's microscopic machinery melts irreversibly. Your veins begin to leak, hemorrhaging blood throughout your body. Meanwhile your intestinal walls are breached, giving the bacteria living in your digestive tract and their toxins access to your innards. There might be vomiting and seizures as you lose consciousness.

Staying cool is as critical to our continued existence as breathing. And evolution's special heat-loss strategy for humans? *Perspiration*.

Sweat as a heat-loss strategy is based on the simple fact that hot surfaces (such as your skin) can evaporate fluids (for example, sweat). As anyone who has tried to simmer down a sauce knows, water evaporation requires heat. In the same way, your body's heat evaporates sweat off your skin.

Because heat is consumed for sweat evaporation, there's a net cooling effect on your skin. It's evolution being crafty with physics and biochemistry: There's heat on your skin and water in your body, so why not dispatch the water to the skin in order to help the body cool down? "Sweat is a great way to dump body heat," says Yana Kamberov, a geneticist at the University of Pennsylvania who is studying the evolution of sweat. "No animal is as good as we are at releasing water on skin and using it to cool down."

But as ingenious as it is, humans are nearly unique in their reliance on sweat to regulate temperature. Most other species use other ways to cool down—some of them unusual and even bizarre. Elephants use their enormous ears to dissipate heat away from their bodies, dogs pant to cool down, and vultures poop on themselves. All of these techniques work to shed excess heat. But none of them work as well as what humans have evolved.

As our ancestors evolved from furry primates into relatively naked, upright creatures, sweating to cool down became one of our species' unique powers. When the weather turned cold, we wore the pelts of other animals to keep warm. But when the weather got hot, sweating was the most efficient way evolution found for keeping us cool.

When our predecessors began evolving body-wide sweat glands

some 35 million years ago, perspiration was as peculiar as it was precious. Many evolutionary biologists include sweating in the portfolio of idiosyncrasies that have helped our species dominate the natural world. "Everybody is very aware of the fact that humans have big brains and that we have language and that we make tools. People should also know and be excited about sweat," says Daniel Lieberman, an evolutionary biologist at Harvard University. "Sweat may seem like an unpleasant problem on the subway but it helped make us human. You can't be physically active the way humans are without sweating. Without sweating, we wouldn't have been able to become hunter-gatherers."

Sweating allowed us to forage out in the sun without overheating, while our predators were relegated to the shade for survival. Meanwhile, other aspects of our unique biology cooperated with our sweat glands to help us stay cool. Bipedalism, for example. Standing on two feet, the hot noonday sun heated only our head, not our entire back and torso, allowing us to trek longer without overheating. When the sun is at its peak, bipedal humans expose only 7% of our body's surface area to the intense heat radiating down. That's one-third less than a similarly sized animal on all fours. And the part still exposed to the sun—our heads—became protectively hairier than the heads of our primate cousins, even as the rest of our skin lost its thick fur.

The fact that humans have a sophisticated cooling-down tool *embedded* in our skin enabled us to run long distances—marathons—without overheating. That means, on the hunt, we could chase our prey to death. Even if our dinner prospects were faster sprinters, our ability to stay cool while running gave us exceptional endurance. Sooner or later our prey had to stop and rest to avoid overheating to death. But humans could keep at it—at a slower pace, perhaps, but over longer stretches. We could force our prey to stay on the move until they collapsed to death from heat exhaustion.

....

The upshot of humanity's cooling-down strategy is a colossal release of salty sweat. One human body has between 2 million and 5 million sweat pores. Collectively, our species' sweat glands number in the quadrillions—more than there are stars in the Milky Way.

In fact, if all the nearly 8 billion people inhabiting our planet collectively stepped into a very hot sauna, we'd produce a combined flux of sweat that could compete with, if not supersede, the water traversing Niagara Falls on a hot summer day. (And yes, a bemused employee at the Niagara Parks Commission helped me figure this out.) That's if we were on the average-to-low end of sweaters. If we were all super sweaters, we'd produce a flux of sweat that would amount to four Niagara Falls.

The millions of microscopic organs responsible for dispatching salty, cooling sweat to skin are called eccrine glands. To me, they look like tiny elongated tubas embedded in skin. There is extensive coiled piping at the eccrine gland's base, which is lodged deep in the dermis. The piping traverses the epidermis to form an exit pore on the exterior of the skin: Imagine sweat leaving the eccrine gland where sound exits a tuba. At the base of these glands, salty liquid is collected from nearby tissues.

As this fluid travels through the eccrine sweat gland to the skin's surface, there's an admirable attempt to salvage some salt. (To cool down we only actually need to evaporate water; the salt is just along for the ride because our interior is a salty ocean.) Yet if you've ever tasted human sweat, you know that the salt-saving machinery is not entirely successful. Working outside for many hours on a sunny, hot day, a heavily sweating person might lose as much as 25 grams of salt—although most of us lose only a fraction of this on a day-to-day basis.

Salty, cooling sweat is one of two kinds of perspiration produced by human bodies. Our other sweaty discharge comes from apocrine glands, which become active at puberty. These are the sweat glands notorious for morphing armpits into stink zones during adolescence.

Apocrine glands are much bigger than eccrine glands, and to me they also look like elongated tubas, albeit tubas on steroids. Apocrine glands sit in the shafts of hair that begin to sprout at puberty, so that their sweaty secretions cover the skin and also coat the surface of these thick hairs. The added surface area provided by hair offers more real estate for the apocrine sweat (and its stink) to blast off into the air. In other words, the hairier your armpits, the more opportunity your stinky sweat has to float up and into the nose of your neighbors.*

When sweat from the eccrine and apocrine glands exits the body, it mixes with oily products of the sebaceous gland, whose job it is to keep our skin moisturized. Although this homemade skin moisturizer is not technically sweat, it often spikes our perspiration with curious, oily chemicals as the fluid travels over our skin.

All of these bodily fluids combine to make sweat a lot more complicated than just salt and water—something the nurse with a predilection for spicy tomato corn chips discovered firsthand. Food and drugs that we swallow, snort, or inhale can percolate out in perspiration. Nicotine, cocaine, garlic odor, food dyes, amphetamines, antibiotics—they all trickle out this route, whether we like it or not. The chip-loving nurse was not the only person to panic when sweat turned colorful: A man obsessed with curing his frequent bladder infections by drinking large volumes of cranberry juice turned his sweat red too, thanks to crimson dyes added to the beverage by the drink manufacturer. Another man fighting recurring constipation swallowed so many laxatives that he turned his sweat yellow, the same color as the ochre dye (called tartrazine) coating the pills.

In addition to cameo appearances from the exotic chemicals

---

* On occasion our bodies also construct hybrid versions of eccrine and apocrine pores, called apoeccrine glands.

present in the food and drugs we binge on, sweat's chemical components also include hundreds of molecules normally found in bodies: waste products from exercise, such as lactic acid and urea, as well as glucose and some metals. Our immune system imbues our perspiration with proteins that do crowd control on the bacteria and fungi living on our skin. These immune molecules help friendly microbes prosper and keep pathogens at bay. And sometimes our perspiration even carries markers of disease, molecular evidence of internal processes, including proteins unique to cancer or diabetes.

Most of the things found in eccrine sweat get into the fluid because they are already circulating in our blood. Eccrine sweat is chemically similar to blood's watery base—it's pretty much blood minus the red blood cells, platelets, and immune cells. Sweat is also chemically similar to the salty liquid keeping our internal tissues moist and hydrated (called interstitial fluid). Most of what comes out in our eccrine sweat is incidental—the chemicals just happened to be coursing around in our blood and then percolated out into our interstitial fluid before our sweat glands got the cool-down directive and dispatched the fluid to our skin.

But some individuals intentionally spike their sweat with chemicals—in the name of science, of course. The idea came to environmental researcher Michael Zech of Dresden's University of Technology when he was sitting in a sauna having an epic, and thoroughly enjoyable, sweat session. Gazing down at the floods of perspiration pouring off his body, Zech wondered how long it would take for a sip of water to transit from his lips to his sweat pores. Unlike many of us who get philosophical in the sauna, Zech has fancy analytical equipment at his disposal.

Before his next sweat, Zech slipped a chemical tracer in his favorite sauna rehydration beverage—a half-and-half mixture of wheat beer and cola. Germans drink this weird concoction in significant enough quantities that it has a name: Cola-Weizen. It's effectively a brown, caffeinated shandy. (Also: Alcohol is permitted and commonly consumed in German spas.)

Zech drank a little over a pint of the tracer-spiked beverage, stripped down, and stepped into the sauna. At regular intervals that he monitored with a stopwatch, Zech captured sweat droplets pouring off his body using little glass vials.

Later, he checked each sweat sample at the lab for the appearance of the tracer. It turned out that it took less than 15 minutes for the tracer to transit through his stomach, be absorbed by the intestine, get filtered through the liver and kidney, enter his bloodstream, lap through his circulatory system to reach the veins in his skin, diffuse through his dermis toward the sweat glands, and then escape out of the millions of pores on his skin.

Question answered, Zech went back to sweating for fun instead of science.

. . . .

Despite being a fundamentally human process, sweat has received a shockingly small amount of attention from researchers throughout history, at least compared to other vital bodily functions. We don't yet know, for example, exactly how many genes are involved in sweat gland production.

Yet the relative dearth of sweat-focused research doesn't mean that perspiration has been entirely ignored by great scientific minds of the past. In the second century, the Greek physician Galen proposed that insensible vapors were being discharged continually from the body, and that this discharge would under certain circumstances be increased so that it took the form of a fluid, namely sweat. Galen also accurately concluded that sweat was sourced from the watery parts of blood.

But Galen was really wrong about sweat too, and his mistakes have seeped into modern-day thinking about the fluid. Instead of realizing that sweat is a sophisticated way to control temperature, Galen thought perspiration was another way to remove waste from the body, similar to other bodily evacuations—feces, urine, menstrual blood, and snot. Galen thought that sweating "cleansed the body of superfluities and of potentially harmful, dangerous, polluting matter," explains Stolberg, the medical historian.

Galen came to this wrong conclusion because of all sorts of intelligent, reasonable observations: That you could, for example, get obese people to lose weight by getting them to regularly run fast, which in turn worked up a sweat. Galen incorrectly deduced that the reason people were losing weight during exercise was because excess fat, dissolved into a liquid, was literally exiting the body through the sweat pores, as opposed to the more complicated reality of calorie and fat burning. Yet Galen's incorrect notion that sweat rids our body of waste is still omnipresent today. Many people rave about the detox benefits of a big sweat. Yet sweat as a detox strategy is as far-fetched as the idea of sweat as a literal fat-leaching strategy.

Certainly, all sorts of chemicals come out in sweat. These chemicals may be toxins, but they may also be useful nutrients or hormones that the body would not want to expunge. Chemicals emerge in sweat because they happen to be floating around in blood, and the human body is inherently leaky—not because sweat is the way your body intentionally expunges toxins. If your body detoxed by sweating, then you would have to expel all 12 pints of blood serum in order to get all the bad stuff out. Doing so would entirely dehydrate you, and you would dry up and die.

Instead, evolution gave us the kidneys—the organs dedicated to filtering toxins out of blood and then directing problematic chemicals into urine and out of the body. Your kidneys are your body's dedicated detoxifiers, so let's leave Galen's two-millennia-old theory in the graveyard of debunked pseudoscience.

After Galen, the science of sweat had a dry spell for about 15 centuries. But then, at the turn of the seventeenth century, a new dawn of perspiration exploration arose thanks to Santorio Santorio, an Italian scientist who was a contemporary of Galileo. Personal measurement was Santorio's obsession. Building upon early work by Galileo, Santorio invented the first device for measuring pulse rates—he called it a pulsilogium.

Had Santorio lived today, he would have absolutely loved the Fitbit. Stuck in the seventeenth century, however, Santorio devised an elaborate hanging chair for measuring changes in his own weight caused by the loss of sweat and other bodily fluids. It was pretty much just a fancy scale. Imagine a thick wooden balance beam. On one side hung an intricately carved chair. On the other side hung counterweights that could be adjusted to accurately measure the weight of Santorio's throne and its occupant. Santorio would spend entire days sitting in his hanging chair, measuring all the changes in weight as he ate and peed and pooped. He did this regularly for 30 years. An obsessive accountant, Santorio showed that what went in did not match what went out: He was losing weight via this mysterious thing called insensible perspiration. Santorio became famous for claiming (accurately) that weight loss from insensible loss of gases (sweat evaporation and breath) surpassed that of all other evacuations combined.

Santorio also became obsessed with consuming portions of food precisely equal to the amount of weight he lost throughout the day. So obsessed, in fact, that he rigged his hanging chair so that as he ate food and gained small amounts of weight, the chair would move away incrementally from the dinner table. When Santorio had eaten the desired amount, the chair would have moved out of reach of the food.

It took a further two centuries before the Czech physiologist Jan Evangelista Purkyně discovered sweat's exit portal in the skin, and in 1833 he announced the existence of these curious eccrine glands. Several decades later, Swiss and German physiologists recorded the electrical signals that are sent from the brain down the spinal cord to open the sweat glands. These neurological impulses—action potentials—give sweat glands the directive to open the floodgates.

And then things got really weird.

In 1928, a clinician in Moscow named Viktor Minor was interested in finding out why some people sweat in varying amounts in varying places, so he developed a whole-body technique for visualizing the onset of sweat. Minor painted the skin of 106 subjects with a solution of iodine, castor oil, and alcohol. Iodine is purplish brown in color, so after the solution dried on the skin, his subjects looked like they were wearing spray-on suntan. Minor then sprinkled his subjects' bodies with starch above the dried iodine, so that they were now a powdery white. When a subject began to sweat, the salty fluid dissolved the dried iodine solution, turning the sweat a brownish purple.

The colored sweat would then percolate past the white starch so that there was a stark contrast between sweaty regions and sweatless regions. Meanwhile, Minor took time-lapse photos of the sweaty progression. His images showed that sweat begins in different places in different people—some people begin to sweat on their faces, others on their torso, others on their legs and buttocks—before the entire body is finally drenched with perspiration.

Minor was bullish about the technique. His subsequent article in the *German Journal of Nerve Medicine* waxes on about the potential of examining the onset of sweat in all sorts of places—the Achilles tendon, for example, or the bald heads of men. He optimistically argued that all manner of neurological and psychological disorders could be diagnosed and studied using his sweat visualization strategy.

Minor's technique spread to Japan, where researchers re-created the starch-over-iodine experiments. These scientists, in particular Nagoya University's Yas Kuno and his colleagues, also figured out how to insert single capillary tubes into individual sweat glands to measure exactly how fast the fluid exited out. (I'm glad I wasn't one of their experimental subjects: They also measured the electrical resistance of different layers of skin by *thrusting*—their term, not mine—electrodes into fingernail mantles, forearms, and the palms of the hands.)

Kuno and colleagues also devised ways to count the number of sweat glands on the human body and reported the average width of eccrine sweat glands—about 70 micrometers, the thickness of a human hair. This figure and many of the currently accepted sweat glands facts found in modern dermatology textbooks—such as that humans have between 2 million and 5 million eccrine pores—date back to work by these researchers of the early twentieth century and Kuno's famous 1934 tome on all things sweat called *The Physiology of Human Perspiration*.

While Kuno was figuring out all there is to know about sweat glands on a microscopic scale in Japan, American scientist Edward Adolph was looking at the bigger picture of how sweat works on a whole-body level, motivated by the entrance of the United States into World War II. When the United States began military operations in North Africa in 1941, top brass wondered how much water soldiers truly needed during desert deployment, particularly during long treks. Adolph, a University of Rochester physiologist, was hired to figure out how much water to supply the infantry to keep its soldiers functional and alive. An optimist might hope that soldiers could avoid lugging around water on long desert marches by deploying a little mind-over-matter. Soldiers could then delay rehydration until arrival at the destination, assuming nobody got lost (which certainly wouldn't happen to an optimist).

This kind of delusional optimism was rife in the US military at the time. Many in the top army ranks thought only wimps needed to hydrate in the desert: Soldiers just needed to buck up, ignore thirst, and get the job done. "Various methods have been tried [for] quelling the thirst sensations, such as chewing, mouthing a pebble or taking drugs. None of these methods is very effective, although they give the man something to do to divert his attention," wrote Adolph in *Physiology of Man in the Desert*.

To settle the matter of thirst, Adolph conducted grueling experiments on soldiers in California's Colorado Desert, a scorching, rocky environment where midday temperatures commonly soar to 110°F—experiments that probably would not have passed muster if faced with a modern-day human research ethics review board.* The soldiers were split into two groups: Half the men had access to water as they marched up to 20 miles for 8 hours; the other half were given none.

This research catalogued many of the first warning signs of heatstroke: high pulse rates and rectal temperatures, thickened blood, gastrointestinal upset, difficulty in muscular movements, and the fact that the dehydrated men "become emotionally unstable," as the scientist put it.

"Only desire to stop and rest," noted Adolph about one man who marched without water for 8 miles in 100°F heat. "Unsociable attitude," he wrote about another. "Began to lag and finally stopped." If any of these soldiers considered desertion, the prospect of more marching in the desert—alone and possibly tracked—wouldn't exactly have sounded much different than their current reality.

Adolph also advised the army on how much water soldiers should be issued on the front lines. Dehydration and rehydration rates depend on individual biology, environmental conditions, the type of clothing worn, and level of activity—variables that the US Army now

---

* Ethics boards that review and oversee research with human subjects were only instituted by Congress in 1973 after the inhumane Tskegee Syphilis Study came to public attention. Sponsored by the US Department of Health, Education and Welfare, the study investigated the effects of untreated syphilis on 400 African American men between 1932 and 1972 in Alabama. The men were deprived of the penicillin cure, even when it became available in 1950.

inputs into fancy computer algorithms to estimate the water needs of soldiers. But back in the 1940s, Adolph gave the army a compendium of average sweat rates, rule-of-thumb figures that are still commonly cited. "A soldier walking at the rate of 3.5 miles per hour in the sun at an air temperature of 100°F loses, on average, 1 quart of water every hour," he wrote. "The same man driving a vehicle under these conditions would lose 3/4 of a quart and sitting in the shade at rest would lose only 1 cup of sweat." Adolph also determined that if you are marching in the desert and you are thirsty and you have some water in tow, it's best to quench your thirst rather than ration: "It is better to have the water inside you than to carry it," he noted.

His work helped show that even if we cannot acclimatize to dehydration, we can acclimatize to heat. After we relocate from a cool environment to a hot one, our bodies begin to adapt by increasing our volume of blood plasma—in effect storing up more internal fluid that can be expelled as sweat. Our sweat rates also begin to increase, so that we sweat sooner and faster than before. We just need a steady supply of water.

In case there are any doubts, heat acclimation works similarly in women as it does in men. Suzanne Schneider, an emeritus professor at the University of New Mexico, says she did PhD research in the early 1970s to help debunk the widely held claim that "men perspire but women only glow." On average there's little evidence to support major sex differences in sweating: Women *tend* to have more sweat glands per unit area and men *tend* to have a higher maximum sweating rate, but many of the sex differences reported can be attributed to other factors such as body size, aerobic capacity, or exercise intensity.*

---

* Of course, the sudden rise in temperature that happens during menopausal heat flashes obviously doesn't affect all human bodies, and it's a fascinating research area. However, the sweating that subsequently ensues after a hot flash is the standard physiological response to an increase in body temperature.

Like other results of military research—drones, canned food, the Internet—Adolph's work fundamentally influenced civilian society. His research informed the still widespread recommendation to cover your skin with long pants and long-sleeved T-shirts in extremely hot and dry environments (which, of course, inhabitants of the Middle East and other desert regions had already known for centuries, if not millennia, before him). The strategy is all about protection from the sun and sweat efficiency, namely how to minimize dehydration. In the hot dry desert, you sweat profusely but the air is so desiccated that water is rapidly whisked off your skin into the parched atmosphere. Your sweat glands, in return, pump out more water to compensate. Loose, long clothing slows down evaporation by creating a slightly more humid environment near the skin's surface. This helps conserve water in an environment where every drip counts toward survival.

In very humid environments, the opposite is true: You want to be nearly naked. That's because the less clothing you wear, the easier it is for sweat to evaporate off your skin and cool you down. At the worst extreme, water already present in humid air blocks sweat evaporation—making it impossible to get heat up and off your skin. You can sweat all you want and you aren't cooled down at all. This is a potentially lethal scenario—and one faced daily by gold miners in South Africa for more than a century.

In 1886, a vast seam of gold was discovered in South Africa's Witwatersrand basin, near Johannesburg, a city that would be founded in the subsequent gold rush. The enormous profits made from this gold and the laws enacted to control its extraction helped establish and uphold apartheid in South Africa. The gold miners—primarily Black men working for extremely little pay—were exposed day in and day out to some of the most severe and dangerous conditions of humid heat stress noted in the twentieth century.

Only a small fraction of the gold stockpile was exposed to the surface. Within a few decades, miners needed to go half a mile underground to extract it. By the 1960s, they were descending nearly 2 miles below the surface of the earth. The mines were hot, extremely humid, and deadly. One Tswana miner described the job as "working in the grave."

Just the trip down into these mines was terrifying. At the beginning of a shift, most miners entered an elevator that could hold 100-plus people, often referred to as "The Cage." The elevator's brake would be released until the cage began to drop at full speed for about half a mile (1 kilometer). As the elevator approached the bottom, the operator would reapply the brake. Sometimes miners would go down a sequence of such elevators until they reached their workday's destination.*

Deep in the belly of these mines, rock fragments containing gold were produced by explosions, with ear-splitting noise that ricocheted off the rock. Temperatures after the blast could exceed 122°F. The rock didn't just hold gold; it also contained silica, which, when blasted apart, produces dust that is extremely harmful to lungs. To reduce the dust hazard, the miners hosed down everything with water. This water wet down the dust and reduced the air temperature from a blistering 122°F to a brutal 95°F—but the water also made the mine's humidity spike. At such high temperatures and humidity, stuck in an enclosed space miles beneath the surface, and doing extremely strenuous manual labor, many men could not cool down enough to survive. In the 1940s, Zulu poet Benedict Wallet Vilakazi penned a protest poem about the manifold dangers of the gold mines called "Ezinkomponi," where he wrote, "The earth will swallow us who burrow . . . all round me, every day, I see men stumble, fall and die."

Workplace safety was poor worldwide for miners in the mid-twentieth century, but especially so for Black miners under South Africa's racist apartheid system. These men faced multiple life-

---

* In more recent times, accidents involving gold mine elevators in South Africa killed 105 miners in 1995 and 9 miners in 2009.

threatening hazards: Exploding rocks. Collapsing tunnels. Respiratory illness. More recently, South Africa's gold mining industry has faced the potential for class action suits from 500,000 miners who contracted fatal lung diseases. Although heatstroke was (and is) just one of many serious health risks faced by these miners, it was still deadly: Hundreds of Witwatersrand gold miners died of heatstroke throughout the twentieth century.

Industry executives didn't want to close the extremely profitable mines. So starting in the late 1920s, the mining industry began hiring or funding medical researchers to figure out why some miners were more at risk of fatally overheating and to develop heat-acclimation strategies for new miners to avoid this. (There's a difference between heat *acclimation* and heat *acclimatization*. Acclimatization happens naturally as you spend time in a hot environment, whereas acclimation is an intentional approach to achieve the same end, typically faster.) Although some of this research helped miners avoid death by heatstroke, it was fundamentally rooted in the economic imperative for continued exploitation of the gold and, by extension, of Black lives. Deaths by heatstroke were reduced—although not entirely eliminated. The research served the mining companies' continued productivity and their bottom line. It is telling that activism from gold miners ultimately played a role in bringing down apartheid.

What we know now is that in extremely hot and humid gold mines, human survival is a tight game of numbers. Even though most of the sweat we produce cannot evaporate off skin in these environments, a tiny fraction often does. It's this tiny, minuscule amount of sweat evaporating off hot, humid bodies that's the difference between life and death. People with larger bodies often have a better chance of surviving these conditions. That's because bigger bodies typically have more skin overall, which means more sweat glands and more surface for cooling. In effect, larger individuals have a higher ratio of skin surface area to core volume, which is an advantage for expunging internal heat.

A variety of acclimation protocols were developed to train miners' bodies to sweat sooner and more profusely. Initially, these protocols involved assigning miners to work in increasingly hot and humid parts of the mine. By the mid-1960s, heat-acclimation protocols looked like this: Over the course of 8 days, large groups of new miners spent 4 hours a day in a hot, humid tent doing step-up exercises on blocks, with periodic rectal temperature checks to monitor whether anyone was at risk for heatstroke. These protocols were, by many accounts and for obvious reasons, widely disliked. By the 1980s, additional strategies beyond heat acclimation were implemented in some mines—such as outfitting workers with vests that contained dry ice.

Half a century later, athletes complete similarly structured (but less intense) acclimation protocols as those of the Witwatersrand gold miners when they train for competitions in scorching or sweltering environments. In particular, athletes acclimate by doing monitored exercise in a hot and/or humid environment over multiple days while their vital signs are tabulated. The protocols are often customized to a particular athlete's sport, the athlete's individual biology, the location of the competition, among other criteria— ultimately to help an athlete acclimate as efficiently as possible but to minimize the misery of that process; for example, current-day protocols suggest athletes exercise intensely in the heat for 60 to 90 minutes at a time, much less than the 240-minute stretch faced by the gold miners.*

---

* As a society, we continue to benefit from early scientific research that lacked ethical oversight and was performed on human subjects who were discriminated against because of racism, ableism, sexism, socioeconomic factors, and much more. This is as true for some aspects of sweat research as it is for other areas of medical science, including anatomy, infectious disease, and psychiatry. The foundation of thermal physiology in tropical environments, as E. S. Sundstroem put it in 1927, "commenced with the inauguration of a colonial policy by the European nations" and continued with investigations focused on establishing "the chances of white settlement in the tropics."

Acclimatizing our bodies to extremely hot environments may become increasingly important as our planet warms. As the sweat scientist Yas Kuno noted presciently in 1956, "sweating is the only process which makes human life comfortable during hot weather and therefore human existence possible in the torrid zone." At some point—if not before climate change makes many parts of our planet inhospitable, then certainly after—we're going to have to learn as a species to appreciate our sweat and, perhaps, to embrace sweating even more than we already do.

....

And yet, appreciate it as we might, sometimes the floodgates open precisely when we *adamantly wish they wouldn't*. Perspiration is frustratingly out of our conscious control. You can hold back other inopportune bodily functions—tears, burps, farts, pee, poop—at least temporarily. Not so for sweat. When our core temperature rises, the information is dispatched unconsciously to our brain's hypothalamus. That's where the executive decision is made to activate the skin's sweat glands, and there's no amount of willpower that can stop this.

Sweat glands don't just open when we're hot: They can also go rogue in moments of anxiety, opening with abandon when we are not even moderately warm. That's thanks to the hormone adrenaline and its sibling noradrenaline, which can kick open both eccrine and apocrine glands. These hormones circulate in our blood when we are sexually aroused, emotional, or just plain stressed. It's possible that in the heady days of our evolution, we were mostly anxious about predators. I'd like to think that stress sweat from the stinky apocrine gland might have evolved to act as a silent but potent odor alert to other humans that it would be wise to start running. Or perhaps stress sweat from the temperature-controlling eccrine gland evolved to be prescient: Our body anticipating the need to sprint away imminently and thus preemptively activating a cool-down strategy.

Even today, the human body anticipates overheating and preemptively activates sweating. The bodies of many elite athletes sweat sooner and more voluminously than those of average people (even when not heat-acclimated), because their physiology has learned—it has been trained—to expect and compensate for high core temperatures caused by long stretches of intense exercise. The same holds true for those who go to the sauna regularly: Spend enough time in hot spaces, and your body opens the floodgates at the first hint of heat.

Yet there's certainly a sweat continuum: Some people just seem to be sweatier than others. Are these differences a quirk of genetics? Do some people have slight DNA alterations that increase their sensitivity to temperature, or their number of sweat glands, or the flow rate of their sweat glands? Could the climate in which someone was born and raised alter their sweating characteristics?

There are arguments for both nature and nurture, says Andrew Best, who is studying the issue with Jason Kamilar at the University of Massachusetts, Amherst. They are measuring the sweat gland density of people around the world to see whether the environment in which they grew up influenced how many of their sweat pores became functional as children, and how many remained so in adulthood. Certainly sweat glands can be *trained* to be more active—that's acclimation—but everybody has a different baseline. Best and Kamilar want to understand what's at the root of the baseline.

Many people who have grown up in cold or temperate climates and then relocate to the tropics as adults find themselves dripping with sweat while longtime locals barely cut a sheen. But you can be sure that people who seem impressively cool and collected—and more specifically, dry—in hot weather are still, most definitely, sweating. It's just that they are doing so very efficiently. Their sweat rate is keyed to release exactly enough perspiration for optimal cooling evaporation, but not so much as to drip.

In fact, the moment anybody starts dripping sweat, that body is being inefficient. It is overreacting and overcompensating, losing valuable internal fluids in the process. Then again, that body might be making a Faustian trade-off by ensuring that in the short term, there's sufficient sweat around to evaporate away dangerous body heat, even at the risk of longer-term dehydration. "You do need to survive *this* moment if you're going to survive the *next*," Best says. "That's when dehydration might play second fiddle to overheating."

....

For all that remains unknown about sweating, we do know one thing: With very few exceptions, everyone is always sweating, at least a little bit.

Water vapor is constantly floating up and off your body. It happens all the time, even when you aren't working out or nervously talking to your secret crush. We mostly notice sweat when it gets extreme, but in reality there's a constant, nearly imperceptible stream of sweat and evaporation taking place throughout the day, as the body makes tiny incremental changes to its internal thermostat. True non-sweaters are exceedingly rare. These individuals have a genetic idiosyncrasy that results in sparse or absent sweat glands. Often they can't survive in the heat without finding some way to spritz water—ersatz sweat—on themselves.

The slow imperceptible flow of sweat is what scientists call insensible perspiration, because it's sweat you don't sense—even though you tend to leave it everywhere. Insensible perspiration is responsible for fingerprints, and the fact that you leave them on anything you touch. And if you've ever found yourself wearing a garbage bag over bare skin—an unusual fashion choice to be sure—it's insensible perspiration that gets trapped beneath the plastic and then condenses unpleasantly back onto your skin. Most of the time it's not possible to see insensible perspiration vaporizing off bodies— unless you try the experiment suggested in the eighteenth century by anatomist Jacobus Benignus Winsløw: "If we look at the shadow of a bare head, on a white wall, in a bright sun-shiny day, and in the summer season, we shall perceive, very distinctly, the shadow of a flying smoke, rising out of the head, and mounting upward."

There's something delightfully poetic about this strategy for visualizing sweat evaporation. But its imagery begs the question: What happens if the evaporating sweat has nowhere to go? What happens if, instead of a garbage bag that can be ripped off when the skin beneath becomes a wet and soppy mess, you are hermetically sealed in? And what if that airtight seal is keeping you alive?

Such was the dilemma facing astronaut Gene Cernan on June 6, 1966, when he embarked on America's second-ever spacewalk as part of the *Gemini 9A* mission. Cernan's goal was to test-drive a propulsion unit outfitted on a backpack while the spacecraft orbited Earth. At the time, NASA engineers didn't realize how physically exhausting it was to move around wearing a spacesuit. This was, after all, only the second-ever spacewalk for NASA astronauts. It took all of Cernan's strength just to put the backpack on outside the spacecraft. His Gemini spacesuit was also extremely stiff, making any movement a struggle, and physical activity in microgravity is tough to begin with.

"Lord, I was tired. My heart was motoring at about 155 beats per minute, I was sweating like a pig, the pickle was a pest, and I had yet to begin any real work," Cernan later wrote in his memoir. (There has, incidentally, been much musing about what Cernan meant by "the pickle was a pest." Some folks think he meant he was in trouble, that is, in a pickle; some think he was referring to a device within the spacesuit; others think he was referring to his penis. Unfortunately, Cernan has passed away and can't be asked to clarify.)

In any case, the 13 pounds of sweat Cernan lost on his spacewalk had nowhere to go given his sealed-up spacesuit. So the evaporating sweat fogged up his visor, making it impossible for the astronaut to see. At the time, NASA had not yet instituted the buddy system for spacewalks, so Cernan was now blind, exhausted, and alone in space. Over the course of 2 hours and 7 minutes, Cernan slowly crawled his way back to the spacecraft and managed to get back inside—alive, but only barely. "I was as weary as I had ever been in my life," Cernan wrote.

Thereafter, NASA improved spacesuits by stitching tubing into undergarments that flushes cool or warm water near an astronaut's skin to keep the individual at a comfortable temperature, and the agency also began spraying anti-fog coatings on the inside of helmets to compensate for any sweaty glitches.

Sweat had finally found its upper limit of utility. Even though this cooling trick of physiology helped our species survive in many challenging environments found on Earth, at 150 miles above the planet's surface, stuck in a spacesuit, sweat was a serious liability.

2

SWEAT LIKE A PIG

I f you think that sweating is a disgusting way to cool down in hot temperatures, consider the alternatives. You could be one of the many animals that rely on other bodily fluids to evaporate away unwanted body heat—think diarrhea, vomit, saliva, and urine.

Take the male South Australian fur seal, *Arctocephalus forsteri*. It lounges around on rocks in the sun hoping the ladies will be impressed by its real estate, a critical factor for finding a mate. These creatures are seriously territorial about their beach-side boulders. No self-respecting male seal wants to share a rock with another guy. But the Australian sun is hot, many rocks do not have shade, and leaving one's rock to take a cooling dip can be risky for the real-estate agenda and quite possibly the mating agenda. In 1973, the biologist Roger Gentry measured the copulation frequency of males who abandoned their landlocked rocks to cool down in the sea: Those that left their rocks on hot days for a swim got half as much sex as the males that stayed put.

This is where urine comes in.

"At high temperatures, land-locked male seals stood on all four flippers and urinated on the rocks, wetting the hair of the belly and the rear flippers. They then lay on one side and extended the wet rear flipper into the air," Gentry wrote. The pee evaporating off the flipper cooled the overheated animal, the same way sweat evaporating off a human arm cools an overheated body. This pee-based cooling strategy has its own scientific term: urohidrosis, *uro* for urine and *hidrosis* for sweating.

Of the bodily fluids we could ostensibly use to evaporate unwanted heat off the surface of our body, surely sweat is preferable to urine? Sweat is certainly better than vomit, the honeybee's favored choice. On a sizzling summer day, a honeybee collecting nectar in a patch of blooming flowers can easily get too hot. Tiny wings working heroically to keep a lumbering body airborne produce a lot of heat—and can make you wonder whether evolution's aeronautics department dropped the ball with these insects. To avoid overheating, bees "regurgitate their stomach contents from the mouth and spread the liquid all over themselves with their forefeet," noted biologist Bernd Heinrich in his delightful book *Why We Run*.

It gets worse for our little flying friends.

Or to be accurate, it gets worse for the overheated bee's hive mates. Honeybees are social creatures and nectar is a valuable commodity—not something to be wasted. So when the bee returns to the hive covered in vomit, its colony mates retrieve the nectar by "lick[ing] off the residual solids that are left after the water has evaporated," according to Heinrich. We could all take a lesson in efficient economization of resources from these insects.

Or consider the case of storks and vultures. To cool down, these birds "poop on their own legs," explained Danielle Levesque, an evolutionary physiologist at the University of Maine. "And then they increase blood circulation to their legs." The blood traveling along the birds' lower extremities is cooled by the evaporation, which reduces overall body temperature by several degrees. As Heinrich put it, "a turkey vulture sitting on a fence post in the sun on a hot day, calmly and deliberately defecating on its naked legs is behaving in a way that makes sense."

Water evaporation—from sweat, vomit, urine, or feces—is by far the most efficient strategy for cooling down in hot weather. Sometimes it happens in coordination with other aspects of biology, such as the curious rete mirabile. Latin for "wonderful net," the rete mirabile is a network of veins that gets pushed up against the skin in response to an animal's high core temperature, bringing circulating blood closer to the body surface. The rete mirabile helps blood traveling from a hot interior to be cooled by contact with air or with areas of the skin that have been cooled by the evaporation of bodily fluids, such as the vulture's skinny, featherless legs.

Animals often aim cooling bodily fluids toward body parts where there is minimal insulation from fat, feathers, or fur, as well as regions that are veiny and skinny. These thin body parts are what scientists call dolichomorphic (*dolicho* for narrow or skinny, and *morphic* for shape). A giraffe is a classic dolichomorphic animal—its body is mostly neck and legs, a figure that helps it optimize cooling in the hot savanna sun. Animals that aren't as svelte as a giraffe need to focus their cooling on their *most* dolichomorphic body parts. For a vulture, it's the legs. A seal pees on its flippers because those stumpy limbs are the least insulated, most narrow part of its body and are rife with veins: Blood rushing through a seal's flipper is optimally cooled by pee evaporation before it returns to the hot interior to bring down the animal's spiking body temperature.

Another cooling advantage of being dolichomorphic—that is, relatively tall, upright, and/or skinny—is that less direct sun hits the body at high noon. So the average giraffe—or bipedal human—does not get as heated by the sun's radiation as, say, a burly wild boar. (The boar avoids the intense, direct heat of the sun by being nocturnal. Some diurnal creatures—animals, who like us, live their lives in the daylight—avoid the intense heat of the noonday sun by changing the color of their skin. For example, in some reptiles, color change increases reflection until the hottest part of the day is over. At that point they go back to darker, more light-absorbent coloring.)

Look at any animal, and it's a reasonable bet that the thinnest part of that animal's physique is involved in temperature regulation. With an elephant, it's the ears. These large, thin appendages have an extensive network of veins. When an elephant overheats, the animal's brain increases circulation to its ears so that the blood rushing through the appendages' thin skin can cool before returning to the hot core. If you happen to take a thermal imaging camera on your next safari, you'll see that an elephant's body is glowing hot while its ears are not, as blood cooled down in its ears works to dampen the temperature in the animal's core.

Bats enter what's called a torpor state to conserve energy as they roost during the day, a physiological phenomenon that also benefits their cooling agenda. Torpor is no midday siesta; it's a stripped-down existence. The animal's bodily functions—what scientists call metabolism—drop down to just 10% normal operation, with only the most essential processes running in the background. Remember that being alive is hot stuff: Just going about your day requires billions of tiny chemical reactions throughout the cells in your body. These chemical reactions create a lot of heat, even if you aren't out foraging or running from predators or trying to hunt. By going into torpor, bats dial down their internal furnace, putting them in a standby mode that reduces their internal temperature production and lets the animals exist in a state of suspended animation. That a mammal can go into hibernation at above-freezing temperatures makes space scientists keen to find out whether humans might also be able to torpor, for, say, a multiyear voyage to Mars or beyond.

But before animals activate their most complicated heat tolerance tricks, most start simple, by altering their situation or behavior. Humans do the same: When hot, we take off a sweater, fan ourselves, or turn on the air-conditioning. Many animals relocate to somewhere cooler: Pigs wallow in mud to keep cool, making the phrase "sweating like a pig" a misnomer. Other animals hide in the shade or in a cave or—if they're small enough—a hole in the ground as a cooling strategy. Or they'll move to a windy place, to let passing air blow their body heat away. Perhaps you've seen squirrels and lizards sprawled out on a cold rock on a hot day. By stretching out, the animals make their bodies as dolichomorphic as possible, and they optimize contact with the cool surface. Overheated koala bears will decamp from their beloved food sources, eucalyptus trees, in favor of hugging wattle trees, whose trunks can be up to 9°C (16°F) cooler than the ambient air temperature. The koala's belly is less furry, less insulated than its back, which is one reason the animal hugs its tree instead of sitting up against it.

....

The efficiency of these behavioral cooling strategies pales in comparison to that of water evaporation. In fact, when air temperature rises above an animal's normal body temperature range, evaporative cooling is the only way to prevent life-threatening overheating. Yet there is a sustainability problem with evaporation: You need a ready supply of drinking water to resupply your body, a resource that's not always available in dry climates. And if you don't have exquisitely evolved sweat glands like humans do, then it can be difficult to control the volume of fluid lost in the process of getting wet.

An animal can get drenched (which is good for cooling) when it uses an explosive evacuation strategy, such as vomiting. But that's it for cooling until the animal can drink some more water and replenish its (precious) bodily fluids. Otherwise it risks dehydration. And it's not easy to vomit *just a little bit* to save some fluid for later. Even with urination, where controlling volume and delivery is a bit more manageable, there's always going to be some wasteful drips that could have been better spent on evaporation.

Which brings us to saliva. Like sweating, licking one's bodily surface is an excellent way to get wet in a coordinated, efficient fashion without wasting too much fluid. For example, one strategy employed by kangaroos is to lick their forearms to kick-start evaporative cooling on those limbs. A kangaroo's forearm has a pretty impressive system of veins, and because that's the skinniest section of its body, it makes sense that the animal focuses its licking there. Saliva evaporation brings down the temperature of the blood rushing by.

Kangaroos use saliva in another way to cool down: By panting, which is yet another form of evaporative cooling. In fact, a lot of furry animals rely on panting to stay cool. That's because fur is often a liability for evaporative cooling. Anyone with hairy armpits or a wet head of hair knows firsthand how fur traps wetness. When evaporation finally does happen, it takes place at the hairy ends first; that is, far away from the veins in the skin carrying the blood that needs to be cooled down.

For animals insulated by fur, the best wet surfaces for evaporative cooling are the nasal passages, tongue, and throat. Think of a dog. When a dog pants, its mouth is wide open and it sticks its slobbery tongue out to get as much wet surface exposed to the air as possible. But as big and wet as a tongue may be, it's still a pretty small surface area for evaporative cooling compared to, say, all the skin on a person's sweaty body. So panting animals compensate for the fact that tongues don't have a large surface area by repeatedly flushing a huge volume of air over that wet tongue to quickly move evaporated water, and thus heat, out of their bodies.

The act of panting pushes evaporated water vapor away from the wet surface so that the entire cooling process can begin anew, every fraction of a second. Imagine if you were really sweaty and somebody turned on an industrial fan right in front of you. You'd cool down much faster because all that evaporated water would blow away fast, making room for more evaporation to happen at your skin's surface.

Or you can think of it this way: When it is super humid outside, such as in a tropical rain forest, the air above your skin is already so saturated with water that not much more can feasibly evaporate, and so you can't benefit as much from evaporative cooling. Conversely, in a desert you don't even feel the evaporation taking place, because water vapor is so quickly whisked away in the dry environment. Panting uses each rapid exhalation to flush tropical rain forest–like humid conditions away from the surface of the tongue so that the incoming, desert-like air can handle a new load of gaseous water evaporating off a slobbery tongue.

Sheep and some other ungulates have fancy panting infrastructure, bony structures called turbinates with particularly intricate architecture. Like a radiator on a car, the sheep's turbinate has layer upon layer of internal structure with narrow air passages in between. And because the turbinate is inside the nose, the surface of these structural layers is wet. So when sheep pant, huge amounts of air are pushed back and forth above the turbinate's interior wet surfaces, which means a lot of evaporative cooling can take place in a tiny space. The air comes into the turbinate dry, and it leaves wet.

The downside of panting is the animal is actively doing something—effectively heaving repeatedly—and that creates its own heat, unlike the passive act of sweating. (Although panting does conserve electrolytes, because the animal loses only water, not salty water.) Another panting hazard is that if it is done improperly, it can mess with the carbon dioxide levels in the lungs.

It's a bit counterintuitive: When an animal pants, there's a risk that too much carbon dioxide might be flushed out of the lungs. A bit of carbon dioxide is needed there—lungs rely on carbon dioxide as a signal that they really should continue breathing. If too much carbon dioxide is kicked out from the lungs, then the breathing circuit falls out of whack; the brain worries about getting too much oxygen, and it tells the lungs to stop breathing, at which point the creature passes out. (This can happen to those who hyperventilate. Luckily, after breathing stops, carbon dioxide levels rise again in the lungs and, in the best-case scenario, eventually reach high enough levels to restart breathing.) To avoid the carbon dioxide conundrum, panting animals typically switch to a special sort of shallow breathing. This way the lungs aren't totally emptied of carbon dioxide with every heave, and the animal still maintains a lot of air traffic over the wet oral and nasal membranes to expedite evaporative cooling.

Many birds also pant, sometimes doing the evaporative cooling over the wet interior of their necks and respiratory system. Other birds, such as pelicans, herons, and cormorants, use a handy pouch below their beaks—the same used to transport fish and to impress potential mates—to carry out gular flutter, a cool-down technique analogous to panting. Gular flutter doesn't create as much internal heat as straight-out panting. It also conserves water and doesn't mess with a bird lung's carbon dioxide levels.

Some birds, like the nighthawk, ostrich, and roadrunner, pant at a fixed frequency no matter how hot they are. Other birds increase the frequency of their panting or gular flutter as their body temperature rises. The tawny frogmouth, a stocky Australian bird often mistaken for an owl, can get its panting as high as 100 breaths per minute at 108.5°F. A very hot hen can pant as often as 400 times per minute.

But if any birds can boast about heat-tolerance skills, it's pigeons and their closely related relatives, doves. Invasive in almost every place on Earth, their cold temperature tolerance is impressive. But the hot-temperature tolerance of desert-dwelling doves and pigeons outstrips that of all other birds. One ornithologist dolefully predicted to me that when climate change takes its ultimate toll, they might be the only birds left alive.

Desert doves are so skilled at heat tolerance that they actually incubate their eggs by cooling them instead of heating them as most other fowl do. That's because doves like to nest in exposed sites, locations that often involve direct sun. Given that desert temperatures can sometimes rise beyond 120°F, and eggs begin to cook at about 104°F, desert doves have to constantly cool their bodies (and the eggs they are incubating) so that their progeny can mature without getting hard-boiled.

One of my favorite studies about a pigeon's ability to air-condition its eggs was published in 1983. Biologists Glenn Walsberg and Katherine Voss-Roberts of Arizona State University went to the Sonoran Desert in the dead of summer, where it can get as hot as 120°F, in order to study the mourning dove's nesting practices. They had to sneak up on the nesting birds during the hottest time of the day to measure the temperature of the birds and their eggs, so that they could then compare the bird measurements with ambient air temperature.

> Before measuring [body temperature] in the afternoon . . . one of us quietly sat for a minimum of 30 minutes at the base of a ladder which had been placed immediately below the nest at dawn. In this position the worker was not visible to the attending parent, which should have recovered from any initial disturbance. After 30 minutes the worker moved carefully up the ladder to within grasping distance of the incubating dove, which usually was reluctant to leave the nest during the hot afternoon. . . . This, admittedly, is a crude and difficult technique; values of [body temperature] were obtained within 1 minute in only nine of 22 attempts.

In other words, the scientists had to lurk in the blistering heat for half an hour, then climb a ladder to grab the bird and stick a thermometer up its cloaca—the hole used for both poop and sex—in order to measure its body temperature. (If only we were all as dedicated to our work as field biologists!)

The scientists found that doves were cooling their bodies (and thus their eggs) by as much as 9°F below ambient air temperature. Since then, ecologist Blair Wolf, at the University of New Mexico, has found that doves and pigeons can cool their bodies and their eggs to a whopping 25°F below ambient air temperature. This is a pretty incredible feat, especially as doves don't typically leave the nest at all during the hottest 8 hours of the day. (Fun feminist fact: It's the male doves that stay in the nest all day to do this egg cooling.)

To be, in effect, miniature air conditioners, these desert doves (and pigeons) can employ a variety of methods to cool their bodies. They pant and they also do something most birds can't: They literally excrete water through their skin. They don't have sweat glands. Instead, the doves' skin gets leaky between cells, and the water inside percolates out and then evaporates away to provide a cooling effect.

A diverse group of animals leak liquid from their skin this way, without having sweat glands. There's the desert cicada *Dicero-procta apache*. This insect dines on juicy plant stems, extracting the liquid interiors and then flushing the fluid out pores in its abdomen and thorax to cool itself by water evaporation. This makes it possible for the cicada to be out and about in the middle of the day when its predators have to hide out in the shade. A diversity of other animals, including kangaroos and multiple species of frogs, also leak a liquid called interstitial fluid from their skin without having sophisticated sweat glands. The bodies of these creatures have recognized that getting a little leaky can save them from life-threatening heatstroke.

....

Eccrine sweat glands—millions of which produce our salty, cooling perspiration—are actually found in all mammals. But in most other animals, the fluid from eccrine glands is not used to cool down, but to provide grip. Most mammals have eccrine pores only in the soles of their feet or hands. In moments of stress, the salty liquid emerging from eccrine glands provides extra friction for landing jumps and for climbing. It's normally only released when the animals are stressed, such as when they need to escape from a predator or catch prey. You can blame your vestigial self when your hands get sweaty under moments of duress: Humans may no longer need to climb trees to deal with (most) potential threats, but our sweaty palms during moments of anxiety reveal that old habits die hard.

At some point in the evolution of primates, eccrine glands began expanding beyond the soles of the feet and palms of the hand to appear on our ancestors' torsos, faces, and limbs. But not in all primates: Baboons, macaques, gorillas, and chimpanzees do have eccrine pores across their bodies. Lemurs, marmosets, and tamarins do not. This ancestral sweaty split probably happened about 35 million years ago. But it's a date with caveats: "Sweat pores don't fossilize," says Jason Kamilar at the University of Massachusetts Amherst. Thus, you can't just look at fossilized specimens from human evolution and say, *Presto, we see a sweat gland!* So researchers looked at which primates have eccrine glands across their whole bodies (Old World monkeys, which scientists call catarrhines) and which do not (New World monkeys, called platyrrhines) in order to figure out the evolutionary pivot point after which perspiration got a biological promotion.

Even so, nonhuman primates aren't enthusiastic sweaters. Although some of our primate cousins use sweat moderately as a cool-down technique, most also rely on other strategies that better serve their hairy bodies: Chimpanzees, one of our closest primate relatives, with whom we share nearly 99% of our genome, rely heavily on panting in hot weather, probably because evaporation as a cooling technique isn't particularly effective on their furry skin.

One of the defining traits of humans is that we are a sweaty, naked ape. "Naked" doesn't actually mean hairless—most of our body fur evolved into very thin hair across the majority of our skin, explains Yana Kamberov, a University of Pennsylvania geneticist who studies the evolution of sweat glands. "We look naked but we are not actually naked—we have the same density of hair follicles as apes have fur follicles." But losing body fur in favor of nearly invisible, miniaturized hair helped our ancestors capitalize on body-wide eccrine glands.

Human skin isn't just a lot less furry than that of our primate cousins; we also have a lot more eccrine glands. "Humans are slightly bigger than chimpanzees but we've got ten times the density of eccrine glands," Kamberov says. It's clear that at some point in our evolution, after our split with chimpanzees about 6 million years ago, our predecessors started losing fur and gaining sweat glands. The question of which came first is a long-standing chicken-or-egg conundrum: Hair doesn't fossilize any more than sweat glands do. So Kamberov began searching for an answer to this question within our genome.

When we are developing as fetuses in utero, our first sweat glands begin forming on our hands and feet in the first trimester. By the halfway mark, 20 weeks, they're developing over our entire body. But the skin's stem cells are fickle—they've got a suite of possible destinies. They might become teeth, mammary glands, hair follicles, or eccrine sweat glands. Kamberov and her colleagues are finding evidence that the biological signals nudging these precursor cells toward an eccrine sweat gland destiny also inhibit the formation of hair.

Once again, evolution appears crafty and efficient. Eccrine glands are most useful for temperature control when there is not a lot of thick hair around. Maybe evolution miniaturized our hair while *simultaneously* dialing up sweat gland production. Kamberov's work suggests the chicken-or-egg conundrum is moot: Instead, evolution may have orchestrated a perspiration two-for-one.

Her preliminary work also suggests that Neanderthals and Denisovans were also sweatier than chimpanzees. I like to imagine our predecessors romping around together, building up a sweat.

....

And yet humans (and other primates) aren't the *only* animal sweaters. Horses also sweat to cool down, but they don't use salty eccrine gland perspiration for this purpose. Instead, they rely on evaporation of sweat from apocrine glands, the ones more commonly associated with chemical communication, stink, and sexual selection. Another commonality: Like humans, horses can sweat from anxiety, thanks to the stress hormone adrenaline. At racetracks, veteran gamblers scour the start line for horses already damp from perspiration. Many gamblers think sweaty horses at a start line is a bad omen—the animal might be too nervous or angry to do well in the race. (When researchers actually analyzed the behavior and appearance of 867 horses in 67 races, they found that pre-race "sweating on its own was not a reliable performance indicator, but in conjunction with other variables might indicate losers.") Scientific analysis of gambling superstitions notwithstanding, horse bodies pre-race may not yet be hot, but the animals know from experience what is coming. Their hormones open the sweat floodgates even before the starting shot.

Cows, camels, and some antelope also sweat from apocrine glands. The thing to remember, says Duncan Mitchell, an early thermoregulation researcher, is that while sweating may be not entirely unique to humans, "we do it so much more and so much better." Humans achieve so much better cooling from sweat than do other mammalian perspirers, he says, in part because we sweat so voluminously.

Consider a cow: At its maximum, a cow produces approximately one-half teaspoon of sweat per minute over a 10-square-foot skin surface. Meanwhile, a heavily sweating human releases more than 6 teaspoons of sweat per minute over the same surface area, about 12 times as much as a cow. More sweat means more evaporation, which means more cooling.

"Another characteristic of virtually all big animals, except humans, is that they will do anything they can to save water," Mitchell says. "Animals will not use water for cooling if they possibly can [because it is often such a valuable commodity in hot climates]. It is a really intriguing question, for which there is no definitive answer, about why humans are very wasteful sweaters."

....

Amid the annals of sweat, there are two oddball entries for two odd-ball animals: the camel and the hippo. One of the frequently cited idiosyncrasies of the hippopotamus is that these enormous herbi-vores produce a reddish-pink sweat that acts as a sunscreen. Sadly, only part of this is true: The beasts do produce a reddish secretion that acts as a sunscreen, as well as a skin moisturizing cream and an antibiotic ointment. But it's not actually sweat, and it doesn't help the animals stay cool in the sun. In order to stay cool, hip-pos hang out in water. They don't need to lose their own water for that purpose.

As for camels: These desert animals are among the strangest and most resilient creatures around. Camels do some sweating from apocrine glands (the ones responsible for stinky armpits in humans). But camel anatomy hosts a cornucopia of additional heat-survival strategies. The fat in their humps, for example, acts as a cooling parasol for their internal organs. Even so, over the course of a hot desert day, camels can let their core body temperature rise by a whopping 10°F—internal heat that is dumped at night when desert temperatures plummet. I'd hate to see any human sustain a fever of 109°F for any period of time.

One body part that does need to stay cool is the camel's brain. But even as the animal's core temperature climbs throughout the day, its nose takes on brain-refrigeration duty. Water evaporating off the moist membranes of the camel's nasal cavity cools nearby blood traversing up into the brain. A few decades ago, scientists even reported that a specialized vein in the camel's face likely acts as "a temperature-sensitive sphincter" to specifically divert cool nasal blood to the brain during heat stress.

Impressive as they are, I don't envy the camel's cool-down strategies—nor those of any other animal for that matter. Even though our voluminous sweating can be annoying at times, I'm particularly relieved that humans don't need to get creative about making the surface of our skin wet with other homemade fluids. Subways around the world would be so much more unpleasant in the summer if we didn't have sweating skills. Imagine hundreds of people cooped up in a hot, confined space with little—if any—access to water. The lucky few who remembered to bring along water bottles would spritz themselves cool to avoid dying from heat stress while the rest of us puked, peed, pooped, and licked ourselves to thermal equilibrium. Instead, we all just sit there miserable and moist from our own perspiration. Given what might have been, sweating is relative bliss.

3

THE SWEET SMELL OF YOU

Annlyse Retiveau is a petite Frenchwoman. When I meet her at Sensory Spectrum, a company in suburban New Jersey that performs odor, flavor, and other sensory analyses for the food and cosmetics industries, she is sporting a perfectly arranged scarf and an air of elegant competence. It's easy to imagine her gently inhaling a fragrance and delivering a brisk, refreshing critique. Except today Retiveau is not evaluating perfume. Today she's going to sniff my armpits.

Down the hall, some of Retiveau's colleagues are assessing a collection of 60 coffee brews from a beverage company looking to launch a new line of roasts. Meanwhile, her boss is working on an odor lexicon for another client's selection of whiskies—words such as *leather*, *vanilla*, *mossy*, and *smoky*. The evaluators will use the lexicon to help categorize and evaluate the malted beverage. Sniffing the armpits of a visiting journalist to show her how deodorants are commonly appraised seems like a lousy assignment in comparison. When I point this out, Retiveau shrugs.

"I am trained to use my nose as an instrument," she says. "It doesn't really matter whether I'm evaluating things somebody may consider pleasant or unpleasant." Among career sniffers, it's actually déclassé—even unprofessional—to complain about unpleasant smells. In this world, people are more likely to criticize simplistic reproductions of complex odors, such as the fake cinnamon flavor used in chewing gum or the faux butter smell added to cheap oil drizzled on movie-theater popcorn. Authentic cinnamon and butter have hundreds of molecules in their odor and flavor profiles; the fake factory facsimiles have just one or two. They are as offensive to odor analysts as dollar-store rosewater is to anyone who has smelled a real rose.

Armpits aren't the only fetid objects Retiveau has sniffed as part of a day's work: Sensory analysts are hired to evaluate whether new diaper materials mask the smell of poop or whether a new line of garbage bags sequesters the stench of rotting food. In any case, sniffing anything all day—even something delightful—can be onerous. Retiveau points out that her colleagues working on those 60 potential coffee brews are unlikely to grab a cup of joe during their break.

And armpit sniffing, it turns out, is a relatively easy gig. Most deodorant companies want sensory analysts to evaluate whether new formulations keep body odor at bay or how one company's deodorant stacks up against a competitor's. It's a straightforward request: How strong is the odor on a 10-point scale? Nobody's asking them to describe and analyze the scents emanating off a subject's armpit. "We don't have to deconstruct the body odor," Retiveau tells me. "We just focus on the overall intensity of the axillae."

Retiveau never uses the word *armpit*. She sticks to the medical term, *axillae*. The word sounds harsh to me—weaponized. ("The stench of his axillae cleaved his enemy in two!") But the origins of axillae couldn't be more peaceful. According to the *Oxford English Dictionary*, axillae is a diminutive of the Latin word for wing. So every time Retiveau uses the word, I begin to imagine body odor shooting out from under armpits on tiny wings.

For most of us, a majority of our body odor originates in the apocrine sweat glands found in armpits. (The other kind, eccrine sweat glands, release the floods of salty fluid when we exercise or are too hot.) Starting in puberty, apocrine glands begin oozing waxy, fatty molecules into the armpit. Although expelled in microscopic quantities and odorless themselves, the waxy molecules are like candy to the millions of bacteria living in armpits, particularly a genus called *Corynebacterium*. As the tiny microorganisms devour and metabolize the greasy apocrine molecules, they produce chemical waste. It's this waste—in effect, bacterial poop—that makes us stink.

Most deodorants contain antiseptics—chemicals that kill bacteria. A deodorant's modus operandi is to annihilate the microbes before they can dine on sweat. Most deodorants also feature fragrances as backup, perfume that aims to distract nearby noses from any stink produced as those microbial populations bounce back as the antiseptic wears off. The products can also contain chemicals that destroy odor molecules. Deodorant companies hire sensory analysts like Retiveau to assess their products so that they can make claims about, say, blocking odor over a specific period of time.

Fortuitously, armpits are ideal testing zones for those who value the scientific method: Everybody's got two, so one armpit gets the product being tested and the second can act as the control. Contrast this to tests for bad breath, where it can be hard for analysts to remember how bad someone's halitosis was before the mouthwash got down to business. For armpits, sensory analysts sniff one pit, pause to let their nose clear, and then slide over to the other pit for comparison.

Humans have a tendency to concoct tools for even the most seemingly simplistic tasks—consider the back scratcher or the electric fork—and armpit sniffers are no exception. Retiveau shows me a sniffing cup. It's a white, conical paper cup, the kind usually dispensed at water coolers, with its pointy end cut off, transforming it into what appears to be an Elizabethan paper-cone collar for a miniature dog. Retiveau places the narrow end in front of her nose. It looks as if she has positioned a mini-megaphone in front of her nostrils. But instead of amplifying sound, the conical shape helps direct armpit odor molecules toward her nostrils when she inhales.

I've got one hand behind my head so that my armpit is open for business. "You want there to be about 10 or 15 centimeters [4 to 6 inches] between the nose and the axillae," she says, leaning in toward me, paper-cone megaphone in place, bending at the waist. Suddenly she takes three unexpectedly dramatic staccato sniffs, which seem so deliberate and absurd that I have to stifle a giggle. I hold my breath and look at the cream-colored walls. I was so excited by the idea that armpit sniffers exist, I neglected to think about whether I actually wanted anyone (let alone an expert nose) sniffing mine. I feel myself start to sweat as Retiveau explains her sniffing protocol.

Every sniff cycle, she says, begins with three to five shallow "bunny sniffs"—yes, bunny sniff *is* the technical term. These sniffs provide an initial indication of odor potency and help assessors distinguish all the odors present. Some judges may add a long inhale at the end of the bunny-sniff sequence to get a fuller experience of the odor. "But I prefer just bunny sniffs," Retiveau says. "If the odor is really intense, and if you take a long sniff, you can get odor irritation really fast. Bunny sniffs help reduce fatigue," she says, adding that judges are allowed to tweak the sniff sequence to suit their nose sensitivity, as long as they are consistent for all the armpits in the trial.

Retiveau is making it very clear that clinical armpit sniffing is not performed willy-nilly: She follows a scientific protocol developed for professional armpit sniffers worldwide that is described in the *Standard Guide for Sensory Evaluation of Axillary Deodorancy*. Most in the field use the conical paper-cup odor megaphone to assess body aroma, and there is a strict routine regarding which armpit is sniffed first.

The subjects—often stay-at-home parents, freelancers, university students, retirees, anyone with a flexible schedule and a desire to make some cash—undergo stringent preselection procedures before their armpits are deemed to be up to snuff. Potential subjects are rejected if they have more than a 20% difference in left-right armpit odor intensity, as are contenders whose armpits are too smelly or not smelly enough. Assessors are taught to rate armpit odor intensity on a 10-point stink scale. To be welcomed into a trial, test subjects can only score between 3.0 and 7.0 on the stink scale. It's a process that reminds me of jury selection: You've got to be balanced, with no extremist tendencies.

The whole assessment process has got to be utterly surreal, I say to Retiveau, who finally cracks a smile. "You often get the same subjects coming back on a regular basis to participate in these studies," she says. "So it actually becomes sort of a social event. Well, an awkward social event."

In fact, life is pretty exacting for those selling their armpits to science. For 7 days prior to the test, subjects can't use deodorants, antiperspirants, antibiotic creams, or any other cosmetic products in their armpits. Also: No swimming. No tennis. No jogging. No armpit shaving. No perfumed substances on the body, including lotion and hairspray. Test subjects can only wear "pre-laundered wearing apparel."

And subjects aren't allowed to clean their own armpits: "Axillae should only be washed at the test site in accordance with a supervised wash procedure," notes the protocol. "Care should be taken not to get the axillae wet during bathing or showering at home." Trained technicians also apply the deodorant with precision, Retiveau says. For example, when conducting tests for aerosol deodorants, technicians use a ruler to ensure that the can is sprayed at a distance of 12 inches from the armpit and so ensure that each armpit receives a defined amount of product. The cans are sometimes weighed in between sprays to determine the exact amount of product dispensed.

I have not been a good subject.

Not only did I apply my own deodorant that morning, but I've also showered, washed my own armpits, used face cream, and recently sucked on a breath mint. When I arrived at Sensory Spectrum, I asked the receptionist if I could use the bathroom, where I put on another preventative layer of deodorant.

Being sniffed by Retiveau—even while wearing a full battery of anti-stink armor—has left me feeling vulnerable and sheepish, as if I've blurted out some awkward personal fact that nobody wants to know. When I mention this, I wonder if Retiveau thinks I'm being hopelessly North American—and I guess I am. "In France, body odor is a lot less taboo," she says delicately, kindly. "Here in the US, consumers are looking for annihilation of any kind of body odor. In France, consumers look for complementarity. Something that pairs with their odor—that doesn't enhance it but masks the bad part and complements the good."

While talking with Retiveau, I've been stealth sniffing myself— you know, a slight tilt of the head that could be misconstrued as a neck stretch, plus a properly timed breath, all to assess whether my deodorant has worn off. I've been relieved to note a strong presence of the citrus-floral odor of my deodorant. Or as Retiveau puts it to me, these odors "dominate."

"You use a product that works well at masking potential odor," Retiveau adds. "Or maybe you're not very stinky to start with."

As I pack up to leave Sensory Spectrum, it occurs to me that I may be offensive to Retiveau's professional nose. I've applied as much deodorant to my body as a teenage boy might apply cologne on date night. I find it comforting, but why? I harbor no delusions that anyone will actually mistake me for a citrus fruit, let alone a flower. Where does our aversion to natural human body odor come from? After all, over most of human evolution, we've prospered by living in close proximity to other humans; shouldn't we be accustomed by now to our aroma? Some researchers have speculated that humans in groups stank so badly that we kept predators at bay. Shouldn't I, then, equate BO with safety? Yet even among my closest network of friends, among the people I feel safest with, our personal aromas are more hidden than many of our deepest secrets.

....

Like many complex scents, body odor can be challenging to characterize—even for experts. In the 1980s and 1990s, sensory analysts began developing fragrance and flavor "wheels" that listed a wide variety of common odor components in complex products such as perfume, wine, and coffee. These wheels are like cheat sheets: They list all the aromatic notes that are commonly found in products. These scents can be affected by a season's weather, the soil characteristics and climate in which a crop grew, or even by the processing done to make the final product (think wine aging in oak barrels or coffee beans being roasted). The wheels have established a set of vocabulary words for discussing these multifaceted aromas, and they help aspiring analysts to identify familiar scents in otherwise intricate odors.

In more recent years, sensory analysts have developed wheels for everything from maple syrup to compost piles. So it should probably come as no surprise that sensory analysts have also picked apart the layers of scent emanating from armpits to build a body-odor wheel featuring components such as grapefruit, goat, wet dog, mint, asparagus, vinegar, cheese, rancid butter, cumin, and onion.

The body-odor wheel's lexicon gives those studying human body odor a language for discussing an individual's unique scent and the aromatic elements that combine to produce that scent. It's where sensory analysis begins to veer toward chemistry. Sensory analysts typically try to describe the components of a person's aroma using recognizable odors, while chemists tabulate lists of specific molecules floating up and out of a given individual's armpits. And some scientists do both.

Until he passed away in early 2020, George Preti was a trim, energetic septuagenarian scientist who hadn't lost his thick Brooklyn accent despite five decades of living in Philadelphia. The first time I met him, he introduced himself as "America's authority on what makes people stink." He was joking but it was also true. When Preti joined Philadelphia's Monell Chemical Senses Center in 1971, he first worked on the chemistry of vaginal odor. "My wife is very fond of telling people that when we first got married, we used to drive around town to women's apartments to pick up tampons in which we had collected vaginal secretions." He had asked her to join him on these collection trips so that women would feel more comfortable donating their secretions to science, and not like they were participating in some creepy scientist's fetish experiment. Soon Preti turned his attention to other fragrant human emissions, including breath and urine, before he found his way to sweat. At his lab, Preti introduced me to his legendary research fridge, which was packed full of bodily fluid samples stored in glass jars, plastic bags, and test tubes. The pungent aroma that emanated from this chilly archive was a testament to his five-decade-long pursuit of human stink.

I was secretly relieved when Preti suggested we go upstairs to the Monell Center's conference room to talk, so we could escape the fridge's odor perimeter. As we settled in, I asked Preti why researchers believe that every person on Earth has a signature body odor: It's impossible, after all, that somebody has tabulated every person's scent in a global stink project.

The strongest evidence that there are no two armpit odors exactly alike, Preti said, comes courtesy of canines. Dogs can often identify individuals with just one sniff of something they've worn. (The exception is identical twins living together and eating the same diet.) It's a technique long used by law enforcement to track missing persons, for good and bad. Using dogs to help find people lost in a forest is clearly a noble pursuit. But during the Cold War, for example, East Germany's Stasi spy agency collected sweat samples of dissidents and other enemies of the state, in case the Stasi needed to track them down with dogs should they go into hiding or try to escape to the West. The use of dogs to search for supposedly criminal BO was employed on the other side of the Berlin Wall too: A West German man was convicted of murder in 1989 on the basis of the body odor he left on the victim's handbag, sniffed out by two German shepherds. As recently as 2007, German police preemptively collected the body odor of left-wing activists whom they suspected might disrupt a G8 meeting held on the country's Baltic coast.

The body-odor-collecting–dog-sniffing strategy has come under fire for all sorts of reasons—most significantly that nobody has proven the technique to be flawless. But many law enforcement officials and many scientists maintain that dogs are exceedingly good at discriminating between two people's body odors. Perhaps there are subtle differences in the chemistry of human body odor that these animals detect with their excellent noses.

A person's inimitable armpit odor most often arises from two primary sources. Every person has an individualized cocktail of molecules released by the apocrine gland.* This collection of greasy, waxy molecules is simply chains of carbon atoms of variable lengths, intermittently decorated by hydrogen and oxygen atoms. The slight differences in the lengths of the carbon chains and the relative abundance of longer- and shorter-chain molecules make for a unique spread of food on offer to armpit bacteria.

Researchers have shown that more *Corynebacterium* present in a person's armpit microbiome means that more sulfurous, stinky molecules will waft off the armpit. But the warm, moist ecosystem of the armpit also attracts other bacterial squatters: Every square centimeter of human skin hosts millions of microorganisms from a variety of genera—think bacteria such as *Staphylococcus*, *Bacillus*, even the yeast *Candida*. Some of the less abundant microbial inhabitants can sometimes play an unexpectedly large role in the aromas wafting out of your armpits. These bacteria, such as members of the genera *Anaerococcus* and *Micrococcus*, can produce molecules that can dominate the symphony of your scent like cymbals do in an orchestra.

---

* And the sebaceous gland, to some extent.

This diverse community's membership is drawn from the places you've lived, your diet, the people you've lived with—even whether you emerged from the womb through the swarms of microbes in your mother's vaginal tract or via a sterile C-section incision. This personalized, motley crew of skin bacteria dining on your unique cocktail of apocrine secretions produces a fragrant bouquet of chemical waste that identifies you as you.

But while each individual's body odor is different, there is something fundamentally recognizable about human stench generally. It's what makes you certain that a stinky human was in the elevator before you, rather than a rank dog or horse. To borrow from the wine-tasting world, there's a "top note" in human body odor. This dominant aroma is present in most of our perspiration; it's a stink we all have. In 1992, Preti and his colleagues discovered that the top note is a molecule called *trans*-3-methyl-2-hexenoic acid; most people describe it as having a rather rancid goat-like stench with a hint of stinky cheese. Soon chemists discovered that *trans*-3-methyl-2-hexenoic acid is part of a whole family of similar goaty-smelling chemicals dominant in human armpits, a chemical family whose members differ only slightly, by a few hydrogen atoms here or oxygen atoms there. There's something delightfully ironic about the fact that what is distinctive about human stench actually smells like an entirely different animal.

Another top note found in armpit aroma is 3-methyl-3-sulfanylhexanol, whose scent is a marriage between ripe tropical fruit and onions. In a study done by the Swiss company Firmenich— the largest privately held flavor-and-scent company in the world— scientists spent 3 years comparing the body odor chemistry of men and women.* (In this case, "a small plastic goblet was used to col-

---

* There's no discussion of trans or nonbinary individuals, and although it is not specified in this study, scientific research has long selectively studied cisgender people. This omission unfortunately exists in most studies on olfaction and body odor.

lect the droplets of sweat that develop from underarm while taking a sauna," notes their scientific paper.) Firmenich researchers found that women tend to liberate more of the tropical fruit–meets-onion chemical (3-methyl-3-sulfanylhexanol) in their sweat, while men tend to produce higher concentrations of the rancid goat odor (*trans*-3-methyl-2-hexenoic acid). That's not to say men don't have elements of onion and tropical fruit in their sweat and women can't reek like a goat; it's just that there is a higher probability that the reverse is true.

In addition to these two top notes, there are hundreds of other scents that make cameo appearances in the body odor of many but not all individuals. Some of these smelly chemicals are not just common in armpits but also in the botanical world. For example, human body odor can contain molecules called alpha-ionone and beta-ionone, which are important to the scent of roses and violets. (She *does* smell like roses!) You can also find eugenol in human body odor, a chemical found in cinnamon, cloves, nutmeg, basil, and bay leaves. And alpha-terpinyl acetate, which has a woody, herbal scent. If anything, human body odor is a potent reminder that nature has an aromatic chemical palette, much like a colorful paint palette, from which it can create a diversity of complex scents that span the gamut from appealing to disgusting.

Among the cornucopia of component odors, human armpits also contain the wild boar pheromones androstenone and androstenol, which remind some people of stale urine and others of a floral bouquet—if they can smell the chemicals at all. By now, it won't surprise you that humans and goats also share several smelly chemicals, including 4-ethyloctanoic acid, a goat pheromone. It's the relative proportions of these common odorous chemicals that stamp a smell as "human" and not goat or wild boar or rose.

To capture the scent of specific individuals, Preti typically asked sweat donors to shower only once daily with odor-free soap and to eschew deodorant for at least 10 days. Once they'd gone through this preparatory phase, deodorant-free donors spent 8 to 10 hours a day wearing textile pads in their armpits that collected apocrine secretions and the odorous remains of their armpit bacteria's feeding frenzy.

Preti extracted the odors trapped in the armpit pads and put them into an analytical machine that separates all the chemicals found in a sweat sample. Once separated, the chemicals were exported one by one through a tube that Preti could sniff. So there he would sit, in front of the machine, sniffing every component in a donor's sweaty armpits, one chemical at a time. This technique, called gas chromatography olfactometry, or GC-olfactometry for short, is one of the best methods scientists have to connect chemical X with odor Y in a complex mixture of aromas. It is how, after many years of searching for that top note in human body odor, Preti was able to identify it as *trans*-3-methyl-2-hexenoic acid.

It turns out that this technique, GC-olfactometry, is commonly used in a variety of industries to find the primary chemical responsible for an odor—the top note. It's how the food industry figures out how to replace expensive ingredients or spices with factory-made mimics. For example, most cinnamon flavor found in processed food is just a factory-made chemical, called cinnamaldehyde, instead of an extract of a cinnamon stick, which contains hundreds of molecules. Likewise black truffle oil is just olive oil with factory-made dimethyl sulfide and 2-methylbutanal. Although these chemicals do form part of the original odor, they are stripped-down versions of the multifaceted original discovered using GC-olfactometry. The technique is also used to figure out the chemical identity of "off" odors in everything from cat litter to canned beans. And of course, to find the components that identify human armpit odor as undeniably human.

It's also thanks to this technique that researchers were able to narrow down which of the chemicals floating off human armpits are actually perceptible to the human nose. Those hundreds of molecules that combine to form our body odor are just a subset of the thousands of molecules that pop off our bodies but which are odorless to us. We may find the world around us to be very smelly, but we don't actually notice the odor of many chemicals floating around in the air, Preti said. This is why, for example, carbon monoxide is so dangerous to us: It's there, it's toxic, but we can't smell it. The same is true for propane, which is why propane manufacturers add in the chemical mercaptan. Humans find mercaptan to be an incredibly unpleasant odor in very minute quantities, which is useful for tipping us off that there might be a gas leak.

Our nasal receptors have a wide range of sensitivity: Sometimes the receptors send strong signals to the brain when just one molecule is present in that headspace. "It's when a little bit of something smells a helluva lot," Preti explained. The receptors in our noses can also send a mere whisper to the brain even when millions of molecules of a certain chemical are present in the air. And no two humans have the same selection of odor receptors in their noses. "Everybody lives in their own sensory world," Preti said. There are about 800 genes in the mammalian genome that code for odor receptors, but humans only employ about 400 of these genes. Not everybody is employing the same set of 400 odor-receptor genes. So even if you think you know your own smell, you may not know how others are experiencing it. And if there's a person whose body odor entices you like no other, it's entirely possible that this person's aroma repels everyone else.

....

Whether we like it or not, body odor is an honest signal. There is nothing in the production or release of body odor that is subject to conscious manipulation. In modern times, we try to mask or thwart the revelation of these secrets using deodorants and antiperspirants, but for most of human history, our body odor was a beacon of truth about our emotions and our health. Which is why I took a trip to Stockholm to visit Mats Olsson, a neuroscientist at the Karolinska Institute.

Olsson's office has an enviable address on a leafy path called Nobel Way. Two decades ago, he started studying human body odor's role in attraction, but he soon grew more interested in whether human aroma delivers inadvertent information—unhappier truths—about ourselves, such as whether we are sick.

Sniffing patients to analyze their body odor has long been a part of medicine. Nurses dressing wounds will sometimes sniff them to check for the sickly-sweet stink of a *Pseudomonas* infection or the fishy aroma when *Proteus* bacteria invade. People with strep throat often have halitosis that smells of poop. There's the now-famous medical sniffer Joy Milne, a nurse who thought the odor of her husband changed when he developed Parkinson's disease. She alerted researchers when she went to a Parkinson's support group and noticed the same odor on other patients. Scientists in Edinburgh put her nose to the test by giving her 12 T-shirts to sniff—half of them were worn by people with Parkinson's, and the others by people without the disease. Milne correctly identified the six people with Parkinson's as well as one of the people in the control group, who didn't know he had Parkinson's but was diagnosed a few months after the experiment. Scientists are now working on ways to detect the airborne molecules that Milne can smell in hopes of developing a new diagnostic tool for the disease.

Some researchers are convinced there's a way to diagnose diseases such as ovarian cancer because dogs can pick out patients from an otherwise healthy lineup. But can normal people, on a day-to-day basis, identify people who are sick? And if so, how early in the disease's progression does our smell start changing? That's when Olsson comes in: He found that people can smell when another person's immune system is activated from a pathogen even before the sick person begins showing symptoms.

In one experiment, he and his colleagues injected eight (consenting) healthy people with a small amount of endotoxin, which is a tiny cellular component found on the surface of pathogens such as diarrhea-causing *Escherichia coli*. When our immune system detects the presence of endotoxin, it activates Code Red and begins launching a counterattack to fight the invading microbe. The test subjects given the endotoxin spent the next 4 hours being monitored in the hospital while their body odor was collected in a tight T-shirt that had nursing pads sewn into the armpits. Thereafter the subjects were released and the T-shirts and armpit nursing pads deep frozen. (Because the test subjects weren't actually infected with real bacteria, just a component, eventually their immune systems calmed down and went back to business as usual.) A month later, Olsson and colleagues got the same folks to come back to the hospital where they were injected with a saline solution and instructed to hang out in a fresh white T-shirt for a few hours.

The scientists then recruited 40 people to smell the T-shirts. Panel members found the body odor of people whose immune systems were on Code Red to be more aversive than healthy BO. The scientists also chemically analyzed the T-shirts to see whether sick individuals just had a more intense odor. Instead they found that many of the folks injected with endotoxin sweated less than did the placebo group. (Remember that when you are sick, the sweats typically only come *after* your fever breaks, not in the first few hours of infection.) In short, the smell panel wasn't responding negatively to the "sick" people's body odor because it was quantitatively more intense; rather, panelists were responding to a chemical cue in the sweat, which signaled that the person's immune system was on overdrive—exactly what happens during infection.

"In the past, the biggest threat to humans was infectious disease," Olsson says. "This has only changed recently," thanks to better hygiene and the discovery of antibiotics and vaccines. It makes sense, he says, that our species would develop ways of using multiple senses to avoid contagious individuals, in this case by finding their body odor unpleasant. "But that this odor cue comes out in sweat so soon after infection with the endotoxin—within a few hours—and that others recoil from these people, that really amazed me." Having a chemical cue that someone's immune system was fighting a battle is like evolution's version of the US State Department's travel warnings: Don't get any closer because there's a war afoot.

....

Researchers have also searched human body odor for cues to our emotional state; in particular, the stink of fear. There's plenty of evidence that humans produce an identifiable odor when we feel fear or anxiety, says Pamela Dalton, a colleague of Preti's at Monell. She has spent a good chunk of her career researching the odor produced in human sweat when we are anxious or afraid. Anecdotally, Dalton says, people who work as interrogators in law enforcement say that stressed-out suspects all start stinking in the same way during questioning, no matter how differently their body odor smelled beforehand. Expressing fear through odor could have been of evolutionary value. In dangerous situations, it might be useful to communicate fear of an imminent threat—say a predator—to others nearby without having to speak out loud.[*]

When researchers want to study the odor of fear, they need to find a way to get study participants to produce samples of stress sweat. This requires a different approach than just getting subjects to exercise or putting them in a sauna—they need to feel so stressed out that they get sweaty. One such strategy is the Trier Social Stress Test, which involves having to prepare a presentation at short notice to be given in front of a group of interviewers. I can see why it's an industry-standard protocol to make people anxious: Subjects have their notes removed moments before they're supposed to deliver a

---

[*] Some evolutionary biologists also postulate that human stink itself was so potent that it may have dissuaded potential predators from attacking.

presentation and are forced to fill up the presentation time regardless of whether they have anything meaningful to say. They are also forced to count backward from 1,022 in units of 13 in front of a group of people and to restart from the beginning any time they falter. Needless to say, the T-shirts of these subjects tend to be soaked with anxiety perspiration.

One patron with enormous interest in understanding anxiety sweat is the US military, which has funded some of Dalton's research. Consider a group of soldiers in a tank: If one soldier feels afraid and starts releasing anxiety-odor chemicals in their sweat, others in the enclosed space may notice and, because fear begets fear, become afraid. That could, in turn, endanger a mission. If researchers can identify the chemical responsible for the stink of fear, there might be some way to capture and sequester it—much in the same way World War I gas masks captured toxic chlorine, phosgene, and mustard gases before they could harm soldiers.

What's fascinating about anxiety odor is that it can sometimes beget fear even when we're not conscious of it. Jasper de Groot and his colleagues at Utrecht University conducted a series of experiments in which they collected sweat from donors who were watching either clips of a horror movie or a BBC documentary about Yellowstone National Park.

Then the scientists connected a panel of women to an electromyography (EMG) machine to monitor the electrical activity in the muscles of their faces, a proxy for their emotional responses (muscles tensing and brows furrowing, for example, show fear or anxiety). These women were then presented with combinations of sweat odor (fearful, neutral) and the videos (fearful, neutral) while the scientists measured the electrical activity of their faces. It turns out that when the video is neutral but the odor is fearful, women still scrunch up their faces as they would if they were afraid. When both visual and odorous stimuli were fearful, the facial reaction was even more severe. What's interesting is that when de Groot hooked both men and women up to the EMG machines and gave both genders fearful and neutral odors to sniff, only the women reproduced the fearful expressions of the odors. The results might be based on the fact that women are better at odor discrimination than men, de Groot says. Others have speculated—controversially—that women, whose physical strength differs from that of men, might have developed a greater sensitivity to odor information as a way to detect danger.

Some law enforcement visionaries have also proposed installing chemical detectors at airports that respond to fear odor. These could alert authorities about an anxious would-be terrorist. In reality, such a device is more likely to buzz continually, given the large number of people at airports who innocently fear flying. The Pentagon tried something similarly ill-considered during the Vietnam War. According to British journalist Tom Mangold's memoir *Splashed!*, the US Army deployed helicopters with a fart detector onboard in order to help locate the enemy. The problem was that the device, which was also designed to sniff out enemy pee and poop, couldn't distinguish between Viet Cong and American odorous evacuations (uh, obviously). According to Mangold, the Viet Cong distributed buffalo urine indiscriminately in the jungle to further confuse the electronic nose.

The military has another vested interest in the scent of fear. If this odor diminishes the resolve or courage of soldiers, perhaps it could be used as a way to instill fear in enemy troops or as a tool for crowd control. Of course, deploying an anxiety-causing chemical agent on enemy troops could be construed as use of a chemical weapon, which would be in contravention of the Chemical Weapons Convention (CWC), to which the United States is a signatory.

Dalton says she has heard some malodors described as psychological weapons, much like a sound or a flashing light, rather than as chemical weapons. She speculates that developers of an anxiety-odor weapon might keep concentrations low enough to avoid irritation of our sensory system so they can frame it as just a psychological weapon instead of a chemical weapon. "It could be seen as a legal loophole," she says. For now, the discussion is moot: Because chemists haven't been able to isolate and identify the anxiety odor yet, there's no way to manufacture and bottle the scent of fear. "But we're getting close," Dalton says.

....

There's another way in which forensic folks are interested in body odor. Eyewitnesses to crimes have long been asked to point out perpetrators in police lineups. But what if the witness didn't see the culprit, yet was close enough to get a whiff of their body odor?

As if being the victim of a crime weren't bad enough, these folks would also be asked to sniff a selection of stinky suspects. Of course, if "nose-witness" identification meant bringing a culprit to justice, a victim might be willing. But can we identify criminal BO? The track record of eyewitness identification isn't exactly inspiring: According to the Innocence Project, a nonprofit criminal justice organization, eyewitness misidentification "is by far the leading cause of wrongful convictions. Nationwide, 75% of wrongful convictions that were overturned by DNA testing involved erroneous identifications from victims or witnesses."

But there's one study that suggests our noses might be reliable, particularly with violent crimes. Researchers in Portugal and Sweden set up a study where participants were shown a video of a violent crime. At the same time, they sniffed a sample of body odor released from a glass jar. A second group of subjects was shown an emotionally neutral video that was also paired with some body aroma. Afterward, the subjects were presented with a lineup of either three, five, or eight odor samples and asked to identify the body odor they had previously smelled.

According to the final report, the body odor associated with men in violent videos "could later be identified in B.O. lineup tests well above chance." While achieving statistical significance makes for good science, it's hard to believe that "well above chance" would cut it in a courtroom, especially when it comes to proving someone is guilty "beyond any reasonable doubt."

But might judges and juries be convinced if analytical technology could match up the precise chemical makeup of a suspect's odor print with odor emanating from objects found at a crime scene, the way that dogs have controversially done in the past?

Devices that achieve similar feats are already omnipresent, from detectors that can identify the odor of explosives to those that can distinguish whether fruits are rotten on the basis of the molecules floating off their flesh. Researchers in Austria, for example, analyzed the body odors of nearly 200 residents of a small Alpine town. When the scientists examined the odors captured in T-shirts, the team discovered they could identify individuals by tracking the levels of 373 chemicals that might be found in body odor.

One can imagine criminals of the future worrying about freshening up the air of a crime scene in addition to cleaning away fingerprints and DNA-containing hairs, tissue, and bodily fluids. But if eliminating body odor traces becomes another task on the criminal to-do list, crooks may want to get their genome sequenced, or at least their ABCC11 gene.

This gene codes for the export machinery of the apocrine gland, which dispatches the waxy, greasy molecules out onto the skin. People with a recessive version of the gene have a malfunctioning export system, which keeps the stinky starting ingredients away from hungry armpit bacteria. Curiously enough, the same machinery is found in the ear and is responsible for injecting yellowish molecules into earwax. The easiest way to assess whether your DNA codes for this feature is to swab your inner ears: If your earwax is white and not yellow, then you probably have the recessive gene and a dysfunctional apocrine exporter. Although the prevalence of this recessive gene is highest among people of East Asian descent, it is sprinkled in DNA worldwide. If you are lucky enough to have two copies of the recessive sequence, your armpits are *relatively* less stinky.

But people with two copies of the recessive gene aren't completely odorless—nobody is. Even when the apocrine transporter doesn't work, some of those waxy molecules that turn into BO can still percolate out, and the oily sebaceous gland adds its own *je ne sais quoi*.

What's incredible to me is that for the most part, the North American mindset is fearful of body odors writ large; we stigmatize it as if we were in the European medieval era, when folks believed that disease was transmitted through malodor. When you stink in America, many people think there's something wrong with you. Yet, complete lack of odor can also be creepy to others. A best-selling book is entirely based on this premise: Jean-Baptiste Grenouille, the protagonist of Patrick Süskind's novel *Perfume*, was born without any body odor whatsoever. He's not only rejected by society but is an evil sociopath to boot. Something about the creepiness of an odorless character pulls on our universal sense of revulsion at the thought of such a person.

When it comes to armpit stink, we're damned if we do and damned if we don't. Or perhaps like most other aspects of human society, it's easiest to fit in if you follow local protocol, from fashion and traffic laws to body odor levels.

Part II

Sweat and Society

4

LOVE STINKS

At Oktyabrskaya metro station in Moscow, a towering bronze statue of Lenin glares along Krymsky Val Boulevard toward Gorky Park. Below Lenin's feet, among the proletariat entourage, a sculpted woman stands with one arm raised in triumphant solidarity, her armpit exposed and victorious. I decide that this is a good omen. I am, after all, en route to a smell-dating event, where Russians will be judging the attractiveness of my armpit aroma.

Billions of dollars are spent every year trying to avoid this exact judgment. For many people, body odor is so unappealing that they mask it with perfumes, deodorants, and antiperspirants. But what if our obsession with blocking BO is interfering with important lines of communication, those helpful messages aroma sends about anxiety, illness, or even romance? When we spray or roll on product could we be blocking our chances of finding love, of finding the person—or perhaps people—who might desire us even more because of our scent?

In this era of swiping left and right in the search for a tryst or a soul mate, smell dating operates on a more analog premise. Instead of swiping, the strategy is wiping; namely, one's perspiration onto a cotton pad. The premise is straightforward: Smell-dating contenders work up a sweat doing high-intensity exercise, their perspiration-rich cotton pads are collected and placed in anonymous containers, and everyone lines up to sniff through the smelly samples. Participants then secretly rate their top preferences and give their picks to organizers, who then reveal the mutual matches. Like the dating app Tinder, a match only occurs when two individuals pick each other's pong.

The only criterion for a romantic match is scent, which is about as logical as any other dating filter. I mean, who cares if you both share a love of, say, taxidermy or Murakami novels? You'll eventually smell the body odor of your lover, and it's probably going to be a make-or-break moment. Smell dating skips to the chase (or, more accurately, it entirely skips the chase) and uses body odor as the first elimination round for mate selection—or date selection, at any rate.

After organizers at a smell-dating event announce the mutual matches, the lucky duo is left to see if looks and personalities also tally up. Typically these dating events have taken place inside a dark venue at night, in the low-lit bars of New York City, London, Rio de Janeiro, Berlin. The attendees at these soirees are clearly self-selective: You've got to have a preexisting appetite for sniffing strangers to motivate yourself off the couch and to a smell-dating event.

Things would be a tad more opportunistic in Moscow. There would be several afternoon and evening smell-dating rounds in the city's most bustling greenspace, Gorky Park, as part of a larger science and technology festival that takes place over a weekend in May. Random people wandering around the park, science nerds attending the festival, and those attracted to the event after seeing it advertised in local media would all participate—or at least that's what Olga Vlad, the event organizer, told me. This being Russia, people who match up with each other at the smell-dating event would be given exclusive entrance bracelets to a nearby VIP lounge tent in the park, so that couples could get to know each other over free, all-you-can-drink vodka cocktails. What could go wrong?

I pause in front of Gorky Park's grandiose carved sandstone entrance to buy pre-scooped ice cream from one of about a dozen bored babushkas. The women are selling their wares from identical gray utilitarian freezer boxes that look as if they have time-warped out of the Soviet Union circa 1975. Cone in hand, I step back to admire the 80-foot-high entrance archway, with monumental columns and carved reliefs depicting the hammer and sickle, as well as baskets brimming with pears, apples, bread, grapes, the bountiful harvest of a successful proletarian state.

Gorky Park is Moscow's answer to New York City's Central Park. Built in 1928, the year after Stalin came to power, the park's entrance gate, like Moscow's marbled subway system, is emblematic of Stalin's vision for public spaces: impressive, rousing, and vastly incongruous with the simplicity that was expected in proletariat homes.

Gorky Park's 300-acre greenspace has long been the site of ice-skating in the winter and romantic strolling in the summer, and its role as a cultural icon is best evidenced by the fact there's both a Russian heavy-metal glam band and a Cold War spy novel named after it. Yet after the Iron Curtain dropped, the greenspace's reputation slipped thanks to neglect, questionable commerce on park grounds, and a handful of rickety amusement park rides long past their prime. Everything changed in 2011 when Moscow's mayor approved a project to modernize Gorky Park with impeccable taste and a multimillion-dollar budget.

Today, mature trees shade people sprawling comfortably in enormous weatherproof beanbag chairs, many working on laptops using the park's free wifi. Couples walk arm in arm through manicured gardens or past groups of people practicing yoga on the lawn. Food trucks offer pulled-pork sandwiches, deep-fried churros, and sushi.

A third of the park is occupied by the art and science festival of which the smell-dating event is just one of many exhibits, all with a (sometimes tenuous) connection to the theme, "the force of attraction." I wander by "Inferno," where a huge line of participants are waiting to put on a robotic exoskeleton as part of a "performance project inspired by the concept of control and the representation of hell." Down a path toward a body of water, a short, charismatic Italian man with a well-coiffed pompadour and a fabulous suit is talking animatedly about his artistic installation to a television camera crew. He has used satellite dishes to make floating metallic swans that emit creepy computer-generated music as they bob around the park's Pioneer Pond. Suddenly we are all distracted by a loud splash as a uniformed soldier jumps into the water. Despite his blood-alcohol level, which is clearly very high, the soldier has the presence of mind to hold onto his hat during the plunge. He emerges dripping, waving his hat triumphantly, and takes a bow while his buddies (also in uniform, also smashed) cheer wildly.

I suddenly realize that there's a multitude of extremely inebriated men mingling with festivalgoers. Wearing identical green uniforms, black boots, and military visor caps, the drunken soldiers are like anachronistic extras with coordinated outfits. And they're everywhere: leaning against trees, lounging on park benches, jokingly attempting downward dog. A woman wearing a festival-organizer badge next to me shakes her head and sighs. "Once a year, these guys come from all across Russia to party in Gorky Park. They are part of the, oh what do you call it . . ." She pulls out her phone to make use of Google Translate. "Ah yes, they are Russian Border Patrol. We did not realize the festival was happening on same weekend as Russian Border Patrol party until it was too late."

Along the shore of Pioneer Pond, a festival organizer with a megaphone is telling folks in English and in Russian to sign up for a smell-dating round. A tall German woman with impossibly straight hair and a friendly smile adds my name to the list, hands me some wet wipes, and instructs me to remove the deodorant in my armpits and any other perfumed products I might have put on today. Mareike Bode is part of an olfactory art collective in Berlin that has been invited by the Russian festival organizers to run several smell-dating sessions over the course of the afternoon and evening.

About 40 people are milling around. A 27-year-old woman named Sofya, wearing a blue bomber jacket and a headband composed of tiny red rosebuds, is surveying the crowd. I ask whether she has ever been attracted to someone on the basis of body odor. "Yes, that's the only way I choose a partner. I prefer that when my partner wears no deodorant, that he smells okay. I have been repelled by a man's body odor." Sofya gives me a significant look that I don't know how to interpret.

"I have a question for *you*," Sofya says. "Have you ever been on the Moscow subway in the summertime? The smell of so many bodies is intense. It's terrible. I think this is the most important problem our government should solve," she adds with an ironic smirk. "But seriously, body odor is important for finding a partner. Well, if we are talking about a serious partner, I prefer to speak first. But when it is about sex, I need to like his smell."

The crowd is primarily composed of people in their twenties and thirties. Alexey, 31, a short, muscular man in a tight white T-shirt, says that a woman's natural smell is "potently important for any relationship. But that's probably because I have a big nose," he said, pointing to his robust aquiline specimen.

Anne Maria is a 21-year-old Italian exchange student who wants to try meeting Russians offline. Sergey and Anya are already a couple. They want to see if they can select each other's scent in the smell-dating game and be matched by organizers. (This, I think, is a perfect way to doom a relationship.) Alek is an impossibly shy, extremely tall 20-year-old with a blond flop of hair who tells me that he really doesn't know if he likes a woman's natural body odor because he doesn't have much experience with dating.

Scrutinizing the crowd with furrowed eyebrows is Dmitri, a swarthy, serious 30-year-old with a fashionably thick beard. He says he eats raw garlic daily for its putative health benefits, as suggested by his mother. "For three years I only use unscented soap for kids and no deodorant. And there's been no difference to my love life," he asserts. When I ask why he eschewed personal care products, Dmitri replies, "Perfumes have made civilization false. Before, when humans lived in small groups, the village community could smell each other. The smell of other humans was a good smell, it meant safety, it was your community."

....

From birth, we rely on our sense of smell to learn the body odor of individuals we love or need the most. A newborn baby, though helpless and immobile, will skootch preferentially toward its own birth mother's odor when breast-milk pads from four different women are placed in the four corners of its cradle. Likewise, a mother can identify her own newborn baby by smell just a few hours after birth. (A parent who didn't give birth can do it too after 72 hours.) Newborn noggins inspire many to inhale deeply. "Family-friendly crack cocaine," is how a friend of mine once described the smell of a baby's head. She's not far off, science-wise. When researchers sampled the body odors from 2-day-old newborns and gave them to women (both mothers and non-mothers) to sniff, the odors activated the reward center of the brain. One wonders whether our brain rewards those who sniff as part of a strategy for helping us learn the odor of our community's newest member.

Sniffing the odor of our loved ones—whether consciously or unconsciously—continues throughout our lives. Siblings and married couples are able to correctly identify the smell of people with whom they cohabitate. Even adult siblings, who haven't seen (or smelled) each other for more than 2 years can still correctly recognize their brother or sister's unique odor print, the signature mixture of chemicals floating off their bodies.

The importance of odor for social cohesion is perhaps best exemplified by the challenges of those who cannot smell. People with

anosmia—the inability to smell—often face relationship challenges: Men without a sense of smell have fewer sexual partners,
while non-smelling women are insecure in their relationships. Both
are more prone to getting depressed. Meanwhile, some research
suggests that empathetic people are more likely to remember the
odor of another person.

Our sniffing abilities and their role in establishing and maintaining social structures can be surprising to some, likely because the
human sense of smell has long been belittled by scholars: The father
of transcendental idealism, Immanuel Kant, thought life would be
better if we all just held our noses so that they were shut off from
the outside world. "Which organic sense is the most ungrateful and
also seems the most dispensable? The sense of smell. It does not pay
to cultivate it or refine it . . . for there are more disgusting objects
than pleasant ones (especially in crowded places), and even when
we come across something fragrant, the pleasure coming from the
sense of smell is fleeting and transient."

Throughout history, many thinkers have argued that vision was a
much more *civilized* way of experiencing the world; using our noses
seemed animalistic, vulgar, backward. If humans sniffed each
other as dogs do, how could we consider ourselves above them? How
could we consider ourselves *enlightened*?

During the 1800s, Western culture morphed its distaste for olfaction into a belief that the human sense of smell was mediocre and
superfluous. To negate the possibility that humans might be unciv-
ilized smellers, we bought into a convenient fib: that the human
sense of smell wasn't very good. More recently, Rutgers University
neurobiologist John McGann penned a fact-check in the prestigious
journal *Science*: "Poor human olfaction is a 19th-century myth."
McGann blamed a nineteenth-century neuroanatomist called Paul
Broca, in particular, for this falsehood. Broca classified humans as
"non-smellers," not as a result of sensory testing but because of his
unsubstantiated belief that the human brain evolved free will at the

expense of our olfactory system. Everybody has seen dogs become so entranced by a smell that they bound off, in seemingly uncontrolled pursuit of some inordinately desirable odorous objective. Surely we're better than that?

Humans are quite fond of our free will, and so you can imagine that we might collectively choose to believe a lie, one that demotes our sense of smell in exchange for increased self-determination. But these two qualities (olfaction and self-determination) are not mutually exclusive. We don't need to nix our noses to be in control of the rest of our body.

And in reality, humans have an excellent sense of smell. Our olfactory bulb, which is responsible for detecting odors, "is actually quite large in absolute terms and contains a similar number of neurons to that of other mammals," McGann has written. "We can detect and discriminate an extraordinary range of odors, we are more sensitive than rodents and dogs for some odors, we are capable of tracking odor trails, and our behavioral and [emotional] states are influenced by our sense of smell."

One of the more delightful proofs that humans can track odor trails is thanks to a group of undergraduates from the University of California, Berkeley. In 2007, a neuroscientist named Noam Sobel, then on faculty there, blindfolded the students, put them in a field, and told them to sniff out a trail of chocolate, like a hound tracks a hare.* Sobel and his colleagues showed that humans (or at least hungry students) can track scents like any other self-respecting mammal and that we do so by making comparisons between the odors floating up into one nostril versus the other.

Being told that there's chocolate to be found would certainly motivate *me* to do some sniffing around. Recognizing, even tracking, the scent of your favorite treat, your brother, your baby, or your

---

* Sobel subsequently moved to the Weizmann Institute of Science in Israel.

lover is not an entirely surprising feat for the nose. It's a matter of correctly remembering the odor of things we've smelled hundreds, maybe thousands of times—the odor of things we've memorized by heart, by ad nauseam exposure.

But there's a big difference between identifying a familiar smell and deducing new information about an unknown person from that person's body odor. Accurately intuiting invisible facts about a stranger on the basis of that person's aroma would require that either we have learned odor X corresponds to characteristic Y or that humans have some sort of inherent, genetically encoded knowledge that odor X corresponds to characteristic Y. Furthermore, deducing anything from somebody else's body odor would require that we lean in and sniff them, an activity that is considered both awkward and creepy in most social circles.

Or is it?

Most human greetings have involved a moment or two of increased proximity wherein we can, at least theoretically, take in the odor of another person. Hugging and cheek kissing are obvious opportunities to sniff each other, especially in parts of the world, such as Europe and the Middle East, where cheek kissing involves pecking back and forth multiple times. (Folks in Corsica greet each other with as many as five consecutive cheek kisses.)

Bowing forward, as the Japanese and Koreans do, also brings two individuals within sniffing distance. And then there's the handshake. It may not bring your nose close to a new person, but it results in a hands-on collection of a new person's sweat and other odors found on the hand—which can then be sniffed later at one's own discretion. At least, all this was true before Covid-19.

Sobel also conducted a fascinating experiment with his graduate student Idan Frumin to see what people did with their hands after a handshake. Their team secretly videotaped people after they shook the hand of someone new, someone they had just met for the first time. Here's their delicious discovery: A few seconds after the handshake, the experimental subjects would inevitably sniff their own hands, to gain some odorous information about the new person.

"When we showed them the videos, many of the subjects were completely shocked and disbelieving," Frumin told me. "Some thought we had doctored the videos—not that we had the computing power or the expertise to do so."

When the new individual was of the same gender,[*] the subjects sniffed their shaking hand twice as much as before. In contrast, after handshakes across different genders, subjects more than doubled the amount of sniffing they did of their own left, non-shaking hand. The scientists speculate that sniffing the hand that contains residues of a person of the same gender could deliver information about potential sexual competitors. In the animal world, many species have as avid an interest in the odors of their sexual competitors as they do in the odors of their potential conquests. "A handshake is a way to transfer that information, and have it, well, in the palm of your hand," Frumin said, "to sniff at your convenience." When Frumin now goes to conferences, he sometimes stands back and watches people unconsciously sniffing. "Sometimes I catch myself doing it too. People tell me I've ruined handshakes for them, that they've become very self-conscious about shaking hands, especially with me."

The scientists speculate that these results are "only the tip of the iceberg." They think that hand-shaking is a sampling strategy humans use to gain aromatic information about one another, information that can tell us useful things about the people we meet.

For example, humans are sometimes good at using our noses to

---

[*] In this and many studies on odor signaling in humans, scientists have not included research subjects who are trans or non binary (or mention if they have). In the (comparatively fewer) studies that include human subjects from across the spectrum of gender and sexuality, there's a tendency to select individuals who fit squarely into gay and lesbian silos.

guess the sex of an unknown person on the basis of the sweat odors that person leaves behind on T-shirts. We can't always do it correctly, but we hit on the right answer enough that some scientists have repeatedly searched sweat for molecules involved in discriminating sex. But although they've picked out a few components in body odor that appear more intensely in men or in women (say the goat versus onion notes), most in the field think that sex discrimination is much more complicated and chaotic than the presence of one or two chemicals.

Many believe that the odors emanating off an individual combine to provide cues about sex, the way that people who have heard a lot of Vivaldi and Bach pieces can often discriminate between their compositions based on auditory cues, even though the music is played from the same palette of notes. Perhaps it is the same for human body odor. That is, after smelling enough people, we learn to discriminate combinations of scents that are more common in men or more common in women, even though the original palette of odors is the same.

Of course our eyes—in combination with limiting assumptions we make in our heads about how men and women look—often suffice to determine gender, but could there be messages in body odor that cue people about more invisible characteristics? In one study from 2005, Monell's George Preti and colleagues asked lesbians, gay men, straight men, and straight women to collect their armpit sweat in cotton pads. They found that the gay males in their study preferred odors from other gay males. But some of the other preferences they tabulated weren't as easy to interpret. When the study subjects were presented with body odors from straight women and lesbians, everyone preferred the odor of straight women, except for straight men, who preferred the odor of lesbians. There's no clear-cut answer about whether our noses help guide us to dates of our preferred orientation, and this study's compartmentalization of sexual preference to gay, lesbian, and straight individuals certainly doesn't provide any insight to any possible links between body odor and whom we desire on the full spectrum of gender and sexuality diversity.

....

These days, we want our life-mates to satisfy our intellectual, emotional, and physical needs. But evolutionarily speaking, for propagation of the species, all humans need is someone with compatible-enough genes that our offspring have a decent chance at survival, at least long enough to procreate themselves, so that our DNA can be passed on to future generations.

The best proof of this comes from a study published by Claus Wedekind in 1995 when he was a graduate student. Wedekind (now on the faculty at the University of Lausanne) showed that women can sniff out a genetically compatible mate—or at least mates with compatible immune systems. Female test subjects were asked to rate the attractiveness of odors emanating off T-shirts worn for 2 days by anonymous men. Meanwhile, blood samples were taken from everyone and their DNA analyzed, specifically a set of immune-system genes called the major histocompatibility complex (MHC).

These genes are involved in helping immune cells learn to recognize foreign, pathogenic invaders. As it turned out, women preferred the odor of men whose MHC genes were different enough that any shared offspring would likely have healthy immune systems.

Infectious disease has been, until modern times, humanity's greatest threat. If you could produce babies with immune systems that were adept at dealing with a variety of pathogens, then your progeny, and your genes, could survive.

The idea is that if you have to fight off future foreign pathogens, whose appearances and vulnerabilities are unknown, it's probably best to have an arsenal of immune weapons that's diverse, so that you can fight a wide variety of microscopic enemies.

At the time Wedekind started his work, researchers already knew that some animals selected mates along these lines. Mice were found to dip their noses into each other's pee to suss out things such as gender and virginity status. On the basis of urine odor, rodents preferentially mated with those who had dissimilar MHC types.

"If you are in danger of mating with a close relative because you don't know them socially, which happens in mice, then a cue that helps you to identify how close you are related to another mouse would help you avoid inbreeding and all the negative consequences that come from it," Wedekind says.

"Humans used to live in small groups over several generations, so there was a certain danger that the few individuals available for mate choice could potentially be relatives. The mother of someone is always obvious, but the father is not always so clear. So there was always a danger of having a baby with your relative. And a cue that would help avoid that would then provide an evolutionary advantage. That could be why we initially developed these preferences," Wedekind explains. "But this doesn't make much sense now," he quickly adds, because the pool of humans to mate with is now enormous, and most people know their family tree. Subsequent research has generally confirmed Wedekind's original work, but attempts to tease out nuances or show that this MHC effect has a major impact in modern human mate choices have been mostly disappointing.*

Yet Wedekind's MHC study is by far the best-known research about body odor and romance. Wedekind says he finds the wide-

---

* There are of course exceptions. Wedekind studied the preferences of women who were on the oral contraceptive pill, which uses hormones to make the body falsely believe it is pregnant. These women had reverse penchants, preferring men with similar MHC genes. The reason why women prefer men with similar immune systems when they are pregnant (or when their bodies are tricked by the Pill into believing they are pregnant) has been the subject of much debate. A common explanation is that a pregnant woman might want to stay close to family who will support her through child-rearing. Men with similar MHC genes are more likely to be genetically related and thus more likely to nurture progeny who share similar genomes—at least in theory.

spread familiarity of his work flattering. "At parties, I'm always introduced as 'the stinky T-shirt guy.'" It wasn't just the general public that took note of the work; in the subsequent decades, hundreds of scientific studies have added what we know about the MHC genes.

But there's still a serious issue with the MHC studies: How do we *smell* these immune-system genes located deep in the nucleus of our cells? You might be tempted to handwave an answer, to propose that the MHC genes code for MHC proteins that emerge in sweat and float up off a body and into another person's nose.

If this were true (and nobody has checked that it is), there's a big problem. These MHC proteins are enormous, much larger than the odor molecules that spontaneously float up and out of our sweat. These MHC proteins are so large and bulky, it's more likely that a hippopotamus would spontaneously evaporate out of a sub-Saharan lake than one of these proteins evaporates off our bodies.

Which is not to say I don't believe the end result of the MHC science in humans: I have no reason to doubt that our sexual predilections might be tuned by the immune systems of our potential mates. But it's important to remember that scientists haven't figured out the details of how precisely that message is communicated. Wedekind himself bemoans the fact that this part is a black box. "It bothers me too that we haven't tracked down the mechanism," he says.

Yet there is other evidence that our perspiration carries messages that might lubricate the path to love and sex. One of the most oft-quoted studies (by the lay public) to this effect took place in a strip club. The scientists, based in New Mexico, wanted to find out if a woman was more attractive to (straight) men during the most fertile period of her menstrual cycle (called estrus) thanks to her body odors, something that's true for many other female mammals.

When the scientists tracked the tips from lap dances and the women's fertility status, they found that the dancers earned the most tips[*] from clients when the dancers were fertile, in the so-called estrus window of their cycles. Although the researchers didn't test their body odor, they argue that something in a lap dancer's scent was communicating to her clients that she was fertile. The dancers presumably wore similar outfits, did the same routines, and had the same personal motivation to earn a good tip throughout the entire month. Yet somehow, the spike in a dancer's luteinizing hormone, which signals the ovaries to release an egg or two, was communicated beyond her body. And this biochemical message was appreciated (one assumes unconsciously) by men who purchased lap dances.

In addition to sweat, another source of smelly chemical information might be tears. In 2011, Sobel (of the chocolate sniffing and post-handshake sniffing experiments) reported in the journal *Science* that he and his research team had collected the tears of women watching a sad movie or an upsetting news report. When men sniffed the women's sorrowful secretions, they had decreased sexual arousal and testosterone levels.

You could argue that simply seeing a woman cry might be enough to clue in her partner that there's not going to be any sex tonight. But the smell might also drive home the message and provide a helpful biochemical dial-down of lust. Sobel told me he'd love to do the reverse study: To see what effect a man's tears had on women, but it was too hard, at the time, to get enough tear samples from men to do the experiment with statistical power. To compensate for this, Sobel is flash-freezing tears of men and women to conserve any chemicals in the tears for later analysis, when enough samples have been collected. It makes you wonder whether prehistoric people cried as

---

[*]   The dancers made about $335 per 5-hour shift during estrus compared to $185 per shift during menstruation. During the rest of their cycles, when the dancers were neither fertile nor menstruating, they earned an average of $260 per shift in tips.

freely as they sweated and what messages might have been present in both those floods.

By definition, body odor messages are fundamentally honest. As Bettina Pause, a psychologist at Heinrich Heine University, puts it, "Neither their production, nor their release, [nor their information] content are subject to conscious manipulation." In intimate situations you have to control your words, your posture, your facial expression, but you can't entirely control your smell. I do appreciate evolution for enforcing a little bit of honesty in matters of romance.

....

"Okay everyone, let's work up a sweat!" On a patch of Gorky Park grass, Alanna Lynch is wielding a megaphone, her dark curly hair held back with a green head scarf. She's wearing a brown tank top, black yoga pants, and bright pink sneakers that flash up and down as she jogs in place to rouse the crowd. The crowd of aspiring smell daters stops milling around and forms a semicircle around her on the lawn. After reminding everyone that they needed to wipe off deodorants, perfumes, and antiperspirants, she cheers, "Let's do this!" and begins to lead everyone through a calisthenic routine "guaranteed to make everyone sweat."

Lynch is an award-winning artist working on an eclectic variety of projects: one involves urine and humanity's response to revulsion; another involves kombucha scobys; she also has a long-term performance piece where she crochets onto herself a body suit using yarn made of her own hair. She considers the sweat-dating event an extension of her work; she and her scent lab collaborators had previously run a well-attended smell-dating event in Berlin. "It was a cold night in November. There were a lot of matches, but maybe there could have been more," she tells me. "I'm not sure everyone worked up enough of a sweat to get a really good odor sample."

Today, Lynch has the weather on her side but she's leaving nothing to chance. In the hot afternoon May sun, she's leading us through a grueling sequence of jumping jacks, burpees, squats, high kicks, and push-ups. By the time we've gone through the first of two exercise rounds, I'm sweating profusely. I see Lynch's collaborator Bode watching me. When our eyes meet, she acknowledges my output with a raised eyebrow and a nod of encouragement, perhaps respect. It's nice to have someone approve of heavy sweating.

At the end of Lynch's routine, Bode hands out little cotton pads. "Make sure you wipe your chest and armpits," Lynch says into the megaphone. Each of us then puts our damp cotton pad into an individual, numbered glass jar. "Don't forget your number," Lynch says. "That's how you'll know if you have a match." I see Alek, the shy young man, sniff his cotton pad before putting it into his jar. "Smells like you?" I ask. He nods. "Definitely," he says with a smirk.

After everyone has handed over the jars containing our sweaty cotton pads, the organizers put the jars on a table. Then there's a swarm as everyone crowds over to sniff the samples, which have not been separated by gender or sexual orientation.

I take a sniff from a jar. It is ripe, metallic, and goaty. Like the odor of a hormonal teenager in the full throes of puberty—plus exercise. I have no desire to sniff that jar ever again. The next jar's scent is barely noticeable, or maybe that's because my nose's odor receptors have gone on strike after the hormonal adolescent pong. I step away and breathe a bit of normal air for a few seconds and then return to the same jar. I think about the professional sniffers who assess armpits for a living and take three short bunny sniffs. I notice a faint smell of onions and grass and earth, like lying in a field on a summer's day. Quite pleasant actually. I note the number: #23.

As I progress through the jars, there's a pretty even split of samples that don't have much of a scent and ones that I can smell but aren't what I'd consider particularly pleasant. Not particularly unpleasant either, but just not all that appealing to me. Some have a serious onion odor, probably female. Others have more goaty top notes, more likely to be male. But who knows, really. One smells like curry. Another like cabbage soup.

I note the numbers of a few samples that I'm on the fence about because I've got to give the organizers a list of my five favorites. And then I hit jar #15. It smells to me like Sex Epitomized. When I sniff again, trying to tease apart the aroma profile, I can detect the standard goaty-oniony background odors of another human, very similar to all the other samples. But something in that mix made me want to sniff again, ASAP. The odor didn't send me into an erotic paroxysm. But it was fundamentally appealing; it triggered an instant reminder that there's this great activity one can engage in with another person, and that it's called sex.

In my reporter's notebook, I wrote down "15!!!!!!!" devoting an entire page to this piece of information. I also wrote #15 on the sheet of paper I returned to the organizers, at the head of my list of favorite odor samples. As I handed in my picks, a sudden twinge of adolescent insecurity passed over me: Might anyone on my list reciprocate? Would I find a match and score a coveted entry bracelet to the VIP cocktail lounge with my potential paramour?

. . . .

It's strong reactions like mine to jar #15 that rouse belief in human sex pheromones, odorous chemicals that catalyze copulation. Insects have them, amphibians have them, mammals have them, so why wouldn't we?

Human pheromones have been dropping suggestive hints that romance is in the air but never quite delivering the goods to the scientists who've been in hot pursuit for decades. Nobody has been able to pluck out a human pheromone from the thousands of molecules floating off human bodies despite enormous effort and a lot of tantalizing indirect evidence. Which doesn't mean they don't exist. It just means that nobody's found the chemical culprits yet, like they have for animals as varied as pigs and moths.

Case in point: bombykol, the first pheromone discovered in silk-worm moths in 1959. Bombykol is a quintessential example of sexual instant gratification. When a female moth has a hankering for romance, all she needs to do is release bombykol in the direction of her desired Romeo and he'll fly over to mate with her. This is *the* definition of a booty call. It works on a vast majority of males, a vast majority of the time.

Another legit pheromone is produced by the male boar, curiously enough, in his saliva. These hairy hogs just need to wander over to a fertile female sow in heat and breathe heavily in her direction. When she gets a whiff of that pheromone, the female spins around, lifts up her rear, and presents it to the male so that he can mount. In wild pig patois, and perhaps more universally, this means "let's start a family."

I shudder to think about what would happen if scientists actually uncovered a chemical with that kind of potency in humans. You don't need much of an imagination to envision the terrible ways that such power would be deployed.

Even if humans produced such a pheromone in our evolutionary past, nowadays its potency has certainly been dialed down by all sorts of competing impulses. We've become highly visual creatures: How a prospective sexual partner looks plays a significant role in whether we try to bed them. And we've evolved agency into our sexual decision-making. Although it can sometimes be a challenge, humans can exert self-control in matters of sex, thanks to some mix of decency, social pressure, and a fear of legal repercussions.

These considerations are unlikely to deter the efficacy of a dyed-in-the-wool real pheromone. For example, that pig pheromone is commonly used by farmers to facilitate artificial insemination. It doesn't matter that there's no male boar around. When the sow sniffs that pheromone, she's motivated to lift her rump for insemination (although some inseminators do rub a sow's hind legs as a boar might to provide a little extra context). This odor–behavior relationship has been encoded into the automatic script of a sow's biology, like breathing and pooping. The female sow is just following her neurobiological imperative.

That's the strict definition of a pheromone. Although scientists have fascinating (albeit esoteric) semantic debates about what constitutes a pheromone, most agree that it is a chemical or mixture of chemicals that consistently elicits the same response in other members of the same species. There is nothing unique in the erotic pull of a sex pheromone. It is anonymously functional on everyone. So even if humans had a sexual pheromone, it would not and could not make you special to your one and only. This is in stark contrast with popular culture's use of the word *pheromone*, summarized in some form of the phrase, "I can't help myself from loving him/her/them; it's his/her/their pheromones." A true sex pheromone, by its strict scientific definition, would certainly make somebody irresistible— to absolutely every member of the opposite sex.

Given that sex pheromones are molecules that typically inspire members of a species to behave like progeny-making automatons, there's a disconnect between the strict concept of a sex pheromone and the subtle evidence that humans are using body odor to learn about and develop preferences for one another. Many scientists working in this field never use the words *human pheromone* when talking about their work. Researchers who have devoted decades of their lives to studying human chemical communication avoid the P-word, opting instead to use words such as *human chemical cues* or *chemical signals* or *social chemosignals*. Because whatever information being transmitted off our bodies through the air and into our noses may be *influencing* our decision making, but it's not *dictating* it.

"We have a problem where we all sort of agree that there's something there in humans but we don't know how to label it," says Johan

Lundström at the Karolinska Institute in Stockholm. "The good thing with [the word] pheromone is people know what you're talking about. You stop someone on the street and they've heard of a pheromone. But it's completely overused commercially. And the general public associates the word pheromone with sexual mating. But very few pheromones [in other animals] have to do with mating. They may assist and inform in mating but they don't induce horniness. The word pheromone has been tainted with sex." And when you know what an honest-to-goodness sex pheromone really does, it's hard to call what we see happening in humans by the same name.

Also there's another problem: the chemistry. Scientists have yet to pluck out important molecules involved in human social communication. Compare this to the case of the silkworm, where scientists know it's the molecule called bombykol. With the pigs, it's two molecules: androstenone and androstenol. These are deemed legit pheromones because they were measured floating off the bodies of these animals and can be shown to consistently change the sexual behavior of the creatures.

In the human experiments, it's clear something or some things (namely chemicals) are floating off one person and being detected in the noses of others. But these floating chemical culprits are still at large. Remember there are hundreds of molecules present in our sweat—and in other bodily fluids too, like tears and earwax, which might also be carriers of information. Although many researchers have tried, none have yet found a molecule in these fluids and convincingly shown that it is, without a doubt, a human pheromone.

That's not to say there haven't been some notable contenders.

For example, the pig pheromones androstenone and androstenol are often found in human sweat, so scientists have tested their ability to change the mood or neurobiology of other humans. Only very slight effects have been seen via brain scans and questionnaires, and only when the human subjects were given extremely high concentrations of the molecules to sniff, at levels that are orders of magnitude higher than that found normally in sweat.

Of course, online peddlers of human pheromone colognes would have you think differently. Products containing the pig pheromone androstenone claim they can help men seduce an unsuspecting woman. The problem is these products are more likely to attract a horny sow rather than a horny human female.

The human pheromone search of the past few decades has involved a lot of impotent effort coupled with attempts to sell unproven products. But many in the field are optimistic a human pheromone will be discovered. At the end of the day, says Tristram Wyatt, an Oxford University evolutionary biologist who has worked extensively on pheromones, humans develop pretty potent body odor at puberty. This is pretty suggestive that we must, in some way, use odor for securing sex.

"I believe we do have some sort of social communication and information transmitted in body odor, whether it's learned or innate," Lundström says. "Whether it's pheromonal or not—that's just semantics. I believe body odor is a complex mixture of compounds that communicates if an individual is sick or not, your age, your gender."

And given that these odors primarily emanate from the armpit, many researchers continue to focus there. As George Preti once said, "When was the last time a guy sniffed your rear end? We are upright creatures and your armpits are near your nose. That's why most researchers looking for human pheromones have studied the armpit."

....

"We've got matches!" Back at the smell-dating event, another organizer is listing jar numbers of people who mutually matched up. I pull out my number from my pocket: #22.

As matching numbers are called out, a motley group of people begin to pair up. Or trio up. Over on my right, Sofya, who thinks Moscow summer subway rides are an affront to the nose, has matched with Dmitri, the bearded hipster who eschews deodorant but eats garlic. And both of them are matched with Marina, a woman in her fifties wearing a pink diaphanous dress, who mentions brightly that she has a cosmic link to the scent of her son-in-law, who may have been her lover in a past life. The three of them start making jokes about being an odd threesome as they stick out their wrists to get VIP bracelets.

"And number 22..." *That's me!* I step forward and hold my breath. "You are matched with number 23!" I look at my notes. Damn. Not number 15, the person who reminded me of pure sex. Ah well, #23 is the person with the delicate body odor, like mown hay, the one that was pleasant, comforting.

I look around. And there she is. Honey-blond hair, hazel eyes, wearing a pair of tight jeans and a cool camel-colored leather jacket. I have been matched with a vision. How did I end up with this babe? She is undeniably gorgeous. Who cares if I don't actually prefer women—I feel like I have *won* the smell-dating contest. I laugh in delight, walk over, and say awkwardly, "I think we are a match? I'm number 22." She flashes me an open, friendly smile and says, "Hi, I'm Anastasia." She tells me that she works in fashion as a handbag importer, and that she writes restaurant reviews on the side. "Wait a second. You're a foodie? Someone with a critical nose and palate picked my BO?" She laughs and we start chatting about the other things we've seen at the science festival. Out of the corner of my eye, a tall gangly guy in his thirties approaches wearing a button-up shirt. His nametag says "Ivan" and he gives Anastasia a look of trepidation. She smiles warmly. "Oh, you're my other match, aren't you?"

Wait, what? *I have competition?*

Anastasia has been matched with both of us, but Ivan and I have only been matched with Anastasia. "Looks like I'll have to fight you for her," I say with a grin, fists raised.

"Okay, fine," he replies. "But let's drink at the VIP lounge first and fight later."

The yellow VIP tent is packed. The structure's white canvas curtain-walls are shaking precariously from the sound system's heavy bass. Dozens of people are propping themselves up against tall white tables. Others recline on couches placed along the VIP tent perimeter. The Italian artist with the pompadour and floating electronic music satellite-dish swans is there, holding court with a crew of enraptured Russians. We head toward the long bar stretching along one side. Bartenders with the chiseled cheekbones of supermodels are serving unlimited vodka berry cocktails.

As my threesome lines up to get drinks, we strike up a conversation with the two guys in front of us from the smell-dating event. I recognize Alexey, the one who believes his big nose helps him appreciate the scent of women. He is standing with Mikhail, a guy with close-cropped brown hair and a gray zip-up cardigan, who looks a little bummed. "It's odd that we matched since we both like women," says Mikhail. "But he seems like an okay guy, so we decided to get the free vodka."

Everyone picks up drinks and heads toward one of the few free tables.

Alek, the tall skinny guy with blond floppy hair and no dating experience, is trying to chat with a short, girl-next-door type wearing a T-shirt printed with hearts and the word *love*. I detour over to them. "We are both students!" Alek exclaims, as though this is a match made in heaven. He is beaming so widely I think he might lose his balance. She also nods enthusiastically. Now *here* is some real chemistry. Not wanting to disrupt it, I smile, raise my glass toward them, and circle back to my own group.

Ivan, my competitor for Anastasia, is talking about his work at a dog shelter in his free time.

"I give up," I say and bow to Ivan. "There's no way I can compete with hobbies as virtuous as that."

"Don't give up too soon," Anastasia says to me flirtatiously. "Well, at least not until my husband comes by to pick me up."

Ivan's face drops. *Dude*, I think, *neither of us ever really had a chance.*

I finish off my cocktail, wish Ivan and Anastasia luck in life and love, and stumble out of the VIP tent in search of food. Walking past Pioneer Pond, I hear an unlikely duet: The snoring of a passed-out border patrol guard mixed with the sounds of the satellite-dish swan song installation. Just then Sergey and Anya, the longtime couple from the smell-dating event, appear on my path, also heading toward the park's exit. "Wait, aren't you going to the VIP lounge?" I ask them.

"We can't. No VIP bracelet because we didn't match. I recognized her scent and picked it, but," and he turns to glare at his girlfriend, "*she* didn't pick *me*!"

5

HOT ROCKS

The burly, hairy man is wearing nothing except a tartan terrycloth kilt and a pair of enormous wraparound sunglasses. As he strides past me through the spa's co-ed changing room, I conclude that we must be heading to the same event: the world sauna theater championship.

I had arrived at a Dutch spa just outside Amsterdam midway through the annual weeklong competition and am thrilled to be in possession of tickets for sold-out performances with titles such as "The Postman" and "Black Swan."

In the changing room, I undress, slip on a bathrobe, stuff the business card–sized tickets into my pocket, and follow Mr. Kilt outside. It's nearing dusk as we walk around a large heated pool near a handful of small wooden saunas that border a forest. The steam rising off the pool's surface is illuminated by underwater lamps, giving the water an ethereal look. Naked people float around with expressions of deep serenity. Suddenly I hear something inconsistent with the bucolic setting: The repetitive thud of heavy bass.

Its source is an enormous circular sauna past several outdoor hot tubs. Some 200 people in various states of undress are waiting in two lines that curve around the building's circumference. I catch sight of Mr. Kilt again and slip behind him in the queue.

Several speakers hanging outside the sauna are pumping out

the kind of jaunty music that typically accompanies politicians as they saunter onstage during election rallies. As we wait in the cool September evening, the crowd bounces up and down to the music to stay warm. Meanwhile, a sauna attendant wearing a microphone headset works the crowd. "Who's louder?" he yells, and people in both lines cheer enthusiastically, competitively.

Finally the attendant opens the sauna door, unfastens the rope holding back my line, and starts taking tickets as we pile in. "No pushing please," he chides.

Bathrobes are thrown quickly on outside hooks, as the crowd rushes in to get prime seats in the circular, Colosseum-shaped sauna. Everyone is naked except for the towels they will sit on— some people try to cover their private zones, others abandon modesty in their rush for the best seats. A few wear Robin Hood–style hats made of felted wool that protect hair from the sauna's searing heat. One woman is sporting a felted hat souped up with Viking horns on top and two red yarn braids on either side. Another guy's felted hat is a proxy for the Norwegian flag—I see a blur of red and blue as he pushes past me. I ditch my bathrobe, dangle my towel so that it vaguely covers my front, and scramble in too.

. . . .

I had been told that sauna theater is a bit like Eurovision, the delightfully cheesy, over-the-top music competition that helped launch the fame of ABBA and Céline Dion. Eurovision is a platform for all kinds of performers—from bearded drag queens and metal musicians dressed as vampires to Russian grannies in full babushka outfits singing catchy pop versions of traditional ditties. Sauna theater features just as broad an array of performers—but often in 185°F heat, with lip-syncing, and in front of naked spectators.

In the center of the sauna arena is an enormous oven, piled high with rocks emanating intense heat. A stage occupies a third of the building's circular perimeter, and it is set up with an entire campsite—including an orange tent. I wonder how its textile fibers are holding up in the blistering heat. To the left side of the stage is a jumbo video screen. The remainder of the sauna's perimeter consists of two seating sections, each with four levels of wooden benches.

The sauna's seating capacity is about 180 people. Many more than that are hoping to squeeze in for the show. A videographer on the second seating level is fiddling with his camera on a tripod. Those in the standby line who don't make it in can still watch the show on a screen in the spa's outdoor restaurant. I find a seat on the first level directly below the camera lens so that I'll avoid its field of view. I have no desire to make any nude cameos for the restaurant diners outside.

Around the room, everyone is arranging and rearranging towels to make space for the last few people squeezing in. The microphoned emcee has instructed us to sit on the wooden benches so that one's entire body is on a towel, including feet. This is so the towel will capture the streams of sweat that will soon slide down everyone's back and legs. Otherwise the combined sweat of the spectators would make enormous puddles on the sauna floor, and people might slip and fall when they leave. Diverting sweat to towels also protects the sauna wood from sweaty salt damage, and it is ultimately more hygienic to keep one's bodily fluids to oneself.

Looking around, I am heartened by the kaleidoscope of flesh settling into place. Bodies of all shapes and sizes sit in anticipation, feet tapping to the energetic music. To keep everyone occupied, the attendant encourages us to do the wave. It is, I hasten to add, impossible to preserve the dignity of your private parts while doing the wave. Even the most beautiful body is awkwardly absurd when its soft sections flop about. But if I opt out, every pair of eyes in the sauna will focus on me and my naked body. *When in Rome* I think as I jump up, convinced that this undulation might go down as the most absurd moment of my life.

....

The awkward but official name for the sauna theater competition is Aufguss WM. The WM stands for *Weltmeisterschaft*, a German word that means "World Championship." *Aufguss* is another German word that means "infusion." To avoid publicly embarrassing yourself at a European spa, you should know that Aufguss is pronounced OW-FFF-goose, as if you wanted to say "Ow! Fuck!" but decided against vulgarity, caught yourself on the "F" and said "goose" instead.

In normal spa circumstances, the Aufguss is a serious, elaborate sauna ritual in which a spa employee called the Aufguss master ceremonially ladles water spiked with essential oils on the sauna's hot rocks to infuse the space with scented steam. As the water hits the rocks, a pulse of hot vapor carrying the scent of clementine, lavender, pine, or eucalyptus envelops everyone in the sauna.

Then comes the towel routine: The Aufguss master begins to whip a towel to create a hot wind that distributes heat in thick gusts around the enclosed space. Like the opposite of a winter windchill, which makes you feel colder, the hot breeze created by the Aufguss master makes the sauna feel hotter. In fact, a good Aufguss master will work up enough wind over the course of the 10- to 15-minute performance that you can actually see your fellow sauna goers' hair blowing in the breeze as the sweat pours down their bodies.

Most Aufguss ceremonies are solemn, relaxing, and venerated, as is the Aufguss master, who is often highly revered. (I've seen Aufguss masters sign the towels of devoted fans after a particularly cathartic session.) If there's music, it's minimalist or New Age-y. People participating in a classical Aufguss ceremony do not do the wave. But this is no typical Aufguss. This is the Aufguss World Championship.

Suddenly the spa attendant slips out and closes the sauna door. The lights go out; the pop music is silenced. As nearly 200 people sit in the darkness, a deep, God-like voice suddenly thunders over the loudspeaker: "SEVEN!"

We all jump. Simultaneously the enormous video screen on stage lights up with a classic film countdown: a black "7" in a circle on top of a white background. God continues to narrate the countdown in time with the visuals. After we reach "ONE!," a new, sexy, conspiratorial voice takes over the microphone and stage-whispers, "SAUNA THEATER!"

As the lights go up, everyone turns to look at the sauna's main door, expecting the grand entrance of the competitor, a Czech man named Jiri Žákovský who will be performing a routine called "The Mountain." But the door does not open. A few awkward seconds pass as everyone looks around confused.

Suddenly there is a rustling from inside the orange tent on stage, and the zipper begins to open. Out crawls Žákovský wearing a snowsuit, boots, and hiking backpack, with sweat pouring down his face and an ice pick in his hand.

The crowd goes wild.

Once the hysterical whooping dies down, Žákovský nods gallantly. He's a twenty-something bearded man with rust-brown hair that's long and braided on top, but shaved below. He looks like he could have been a wildling actor from *Game of Thrones*. Žákovský welcomes everyone to the show and proceeds with a few housekeeping items that are required to get full points from the Aufguss WM judges sitting in the audience: He tells us the performance will last about 13 minutes (there are demerit points if the show lasts more than 15 minutes or less than 10). And he reminds us to shower off the sweat we'd soon accumulate before jumping into any of the spa's pools. And finally he assures us it is okay to leave the sauna at any point during the Aufguss performance if anyone starts feeling woozy from the heat. I can't believe he's not already woozy himself, what with the winter getup.

Then his face darkens dramatically. He tells us that his brother disappeared the previous year while attempting to summit a mountain. Žákovský explains that he wants to traverse the same route, to conquer the deadly peak so as to honor his brother. Photographs of the two brothers wearing winter toques flash on the jumbo video screen, along with images of spectacular winter mountain landscapes. As he tells the audience this backstory, Žákovský trudges around the sauna as if he is hiking up a steep incline. Suddenly he bends over and picks up from the sauna floor the very same winter toque worn by his brother in the on-screen photos—a sad sign that his brother has most likely died on the mountain. With a look of agony, Žákovský removes his own cap, throws it away, and puts on his brother's. As a heavy-metal ballad winds up on the sound system, Žákovský vows to take revenge on the mountain that stole his only sibling.

Next he gets to work, setting conical, mountain-shaped snowballs infused with chamomile essential oil on the hot sauna rocks. A pulse of steam from the melting snow sends the heady floral scent wafting through the room, giving us the sense of being amid a sea of mountain wildflowers. He pours more water on the stove's hot rocks, creating more bursts of fragrant steam around the sauna.

And that's when the aerobic show begins.

Žákovský picks up a thick white towel and begins whipping it around in a helicopter-type motion. The spinning towel spreads the steam and the chamomile scent around the sauna so that the spectators experience a steamy wind blowing over their skin. It reminds me of a gymnastic ribbon routine, but with a towel instead.

The sauna was already extremely warm before he started pouring water on the rocks. Now all these extra gusts of wind are bringing that hot steam in contact with my skin, making me feel even hotter than before.

It seems counterintuitive, but the human body is one of the coolest objects in a hot sauna. Even though your skin temperature in a sauna rises to a few degrees above normal—to about 109°F—the rest of the space is typically about 175°F to 195°F, and the steam is over 210°F. Because your body is relatively cool by comparison, all that evaporated water swirling around condenses on your skin like kettle steam on a cold winter window.

Condensation is inherently an exothermic reaction, which means it releases heat—right there on your body. It's the opposite of what your body is doing when it tries to cool down with sweat: You sweat so that its evaporation off your body has a cooling effect. Conversely, condensation of a sauna's hot steam onto your body heats you up. So when a sauna steams up and the condensation on your skin heats you up, you sweat even more to compensate.

It feels like my pores have opened up and every iota of liquid in my body is trying to escape. The sweat mixes with the condensed steam on my skin to create a flood of liquid rushing down my body, soaking my towel. I notice a pool of fluid dripping off my elbow onto my neighbor's towel. I try to rearrange my appendages in order to keep my sweat to myself, but it's a bit of a lost cause. My slight adjustment now means that my cross-legged foot is dripping on the towel of my other neighbor. Neither neighbor pays any attention to me—the show is too enthralling—so I decide not to sweat my sweat.

The fluid loss feels incredibly purifying, but also alarming. I begin to wonder whether my body will ever close the floodgates or whether I might desiccate right there on the sauna bench. German scientists, it turns out, wondered the same thing and conducted studies in 2015 to find out what fraction of the liquid pouring off a body during an Aufguss is sweat and how much is condensed water. They established that between 30% and 55% of the liquid rolling down your body is condensed water, and the rest is sweat. The exact ratio depends on your personalized sweat rate and the temperature and humidity conditions of the sauna. But knowing this is only partially comforting. There are still cups of water from inside my body being dramatically expunged.

Meanwhile, Žákovský is working the room. Grabbing the corners of his towel in each hand, he whips the textile up and down in coordinated, graceful strokes that play with the physics of the sauna. Even in the hottest of spaces, heat rises; sometimes the difference in temperature from floor to ceiling in a sauna can be 10 degrees or even more. By whipping the towel up and down, the Aufguss competition contender is pushing the hottest air at the top of the sauna downward and the cool air upward. The switch between hotter and slightly cooler gusts gives me goosebumps and the (pleasant) sensation of having a fever.

When a lull in the music finally arrives, I wonder if the performance is over and whether I can now escape the heat. But it's just the end of the first of three acts. My towel is already drenched with sweat, as is Žákovský—or at least what little I can see of his skin, given the elaborate costume.

In the story's second act, Žákovský finds himself in over his head—literally—as dry ice engulfs him in a theatrical avalanche he has concocted on stage. As he escapes the deluge of snow, the smell of mountain pine spreads through the sauna. He has placed snowballs infused with pine essential oil on the hot rocks. Soon Žákovský embarks on more complicated, showy towel maneuvers: He tosses the towel in the air, pizza chef style, and effortlessly catches it again. Then he spins the towel behind his back, lets it fly up into the air, and catches it with the other hand—all while being heavily laden down with trekking gear, including a backpack and ice pick.

Suddenly on the loudspeaker, Žákovský's brother begins to speak from the afterlife. The disembodied voice tries to knock some sense into his sibling and begs him to abandon the venture by arguing that the trek amounts to suicide. Žákovský turns to his towel to work out this internal struggle. Now instead of just whipping one towel around, he does simultaneous throwing tricks with two towels—one in each hand. I looked around the room. Many of Žákovský's Aufguss competitors are gazing intently at his tricks with a mixture of awe and reverse-engineering. Everyone is cheering, even his opponents. One guy keeps jumping up off the sauna bench in euphoria.

Finally, Žákovský capitulates. Realizing he needs to take his dead brother's advice, Žákovský decides to head down the mountain before it takes his life too. To convey his new lust for life, Žákovský infuses the sauna with the scent of a rare flower that only grows in certain parts of the Himalayas.

And that's when Žákovský begins smiling. Certainly it's part of the evolution of his character, but he also knows he is nailing the performance. Žákovský begins nodding at friends in the audience, all while effortlessly landing towel tricks. As far as I can tell, he has not made a single mistake. The music ends, he takes a bow, and utter chaos ensues as people crowd around congratulating him. As everyone files out of the sauna, one air-punching enthusiast keeps yelling "Bravo!" while a Dutch couple whisper to me that they think he's a contender for first place. Outside I see Mr. Kilt raving to a group of spectators in Italian.

I slip away and douse myself with refreshing cold water in a long, slate shower room with enough nozzles for a dozen people. I feel fabulous, exhilarated. As if I have just accomplished something physically challenging, and won.

....

Sauna euphoria is due to some pleasant brain biochemistry and basic physiology: When your skin temperature spikes in a sauna, so does your pulse. By the time you emerge from a 10- to 15-minute stint inside, your heart is probably beating about 120 to 150 beats per second. For many people, this is the equivalent of mild exercise. Meanwhile, your body has been enduring mild heat-shock, with beneficial downstream effects on your blood chemistry. Sauna sessions boost blood levels of epinephrine, growth hormone, and endorphins—the latter of which are, incidentally, also hormones often held responsible (although controversially so) for a runner's high. With a sauna, you get the happiness without the mileage.

As it turns out, there are other parallels between exercise and sauna-ing. Both are beneficial to your heart and cardiovascular system. But let's be clear: Spas worldwide make all sorts of claims about the health benefits of going to the sauna—and most are utter hogwash. Going to the sauna does not cure cancer. Sauna-ing is *not* a smart chemical detox strategy; in fact, it's not a detox strategy at all. Having a good sweat can certainly rid you of toxic emotions, thanks to euphoria hormones that get released into your bloodstream. But calling a sauna session a "detox strategy" reveals a fundamental misunderstanding about how the human body works.

Your kidneys are responsible for detoxification; they are the organ devoted to ridding your blood of chemical contaminants. Certainly all sorts of interesting things emerge in sweat, including chemicals your body would like to expel, such as some heavy metals, cocaine, and NikNaks Spicy Tomato red flavoring. But the fact that those things appear in sweat is incidental, purely because your body is leaky. In order to cool down, your sweat glands siphon water out of your blood; any chemicals floating around in your circulatory system are just coming along for the ride.

A lot of dodgy spa claims of health benefits come from studies that were done decades ago, often with no control group. Or the studies used sample sizes that were too small to prove anything statistically or supply confidence in the conclusions. For example, there's an oft-repeated claim that going to the sauna boosts your immune system and decreases the incidence of colds in winter. The evidence for this comes from a handful of studies from the 1970s and 1980s that even a proponent of the research called "mostly retrospective and poorly controlled." This same German researcher set out in 1989 to do a more serious study. He picked 50 people and put half of them in a sauna-going group and half in a non-sauna-going group. For

6 months the sauna-going folks went an average of 26 times to the sauna, or about once a week; they ended up catching 33 colds, while the non-sauna-going folks caught 46 colds. These numbers are suggestive that maybe there could be something to the idea, but "the mean duration and average severity of common colds did not differ significantly between the groups. It is concluded that regular sauna bathing probably reduces the incidence of common colds, but further studies are needed to prove this."

Many spa entrepreneurs point to this small, 30-year-old study as proof that their establishment helps you sidestep winter's most common malaise. It is certainly possible that future scientists might deliver conclusive evidence that regular sweating boosts immune-cell regeneration (perhaps because you often sleep better after a sauna session and immunity is built during slumber). But there is no such conclusive evidence today.

Going to the sauna has, however, been shown to be excellent for your heart. This conclusion is based on a large study that has been ongoing since the mid-1980s and that is investigating risk factors for cardiovascular disease among Finnish men. The men who went to the sauna regularly had a lower chance of sudden cardiac death, fatal coronary heart disease, fatal cardiovascular disease, and lower all-cause mortality, which means, in short, that going to the sauna regularly could help extend one's life.

Of course, for Finnish men, "going to the sauna regularly" means *more than four times a week*. That is only viable if you have a sauna in your house, as many Finns have. In fact, only 12 of the 2,327 men in the entire study didn't go to the sauna at all. Given the widespread cultural habit of regular sweating, the scientists used the men who went to the sauna once a week as a base reference and compared their health to the men who went more often. As it turns out, the risk of dying of cardiovascular disease was 27% lower for men who were in the sauna 2 to 3 times a week and 50% lower for men who were in the sauna 4 to 7 times a week compared with the risk for the men who indulged just once per week.

So why would going to the sauna be good for the heart?

The heat in a sauna causes blood vessels to dilate, or to widen, which allows more blood to travel through your circulatory system. Wider veins means it takes less thrust from your heart to push your blood around your body, so your blood pressure drops. Low blood pressure is certainly a perk—as long as it doesn't drop too low.

As your circulatory system ramps up its activity (thanks to wider veins), a larger flux of blood reaches the surface of the skin. Sweat glands skim from the blood's fluid base and send that liquid to the skin's surface. The sweat evaporation cools down the skin's surface, which then cools down the blood flowing by. This newly cooled blood travels to hotter internal areas, such as your brain, which cannot afford to overheat.

Wider veins also means there is more blood flow to the surface of your skin, which is partly why your skin reddens when you are hot. It's a good sign that there is more blood rushing to the skin surface to fight the spike in temperature.

Even though you are, in principle, relaxing in the sauna, your circulatory system is not. It's on full-blast. So your heart is experiencing something analogous to a workout, but without the full-body physical exertion. All that blood pumping through the system has knock-on biochemical effects that likely activate plaque-clearing and other benefits to your circulatory system. I say "likely" because the most convincing research into the connection between sauna and beneficial blood biochemistry has not been done in humans; it has been done in hamsters. Researchers in Japan put hamsters into infrared saunas to study the health effects of heat on their tiny cardiovascular systems.

Thanks to their work, we know that heat exposure activates production of an enzyme called nitric oxide synthase 3 in the cells that line the interior of veins and arteries (of hamsters). The enzyme spurs the production of nitric oxide, which is the marriage between one atom of nitrogen and one atom of oxygen. This simple molecule is at the apex of cardiovascular health: It helps prevent plaque buildup in blood vessels; it induces the proliferation of smooth muscle cells and decreases blood pressure; and it helps immune cells called platelets do their job in the blood. Not bad for a trip to the sauna, right?

But don't cancel your gym membership yet. Even though you can get similar heart benefits, don't think you can ditch exercise and replace it with sauna time. Sitting in a sauna doesn't burn nearly as many calories as a workout, and you don't build or strengthen muscle. And the promising evidence for cardiac health benefits was found for people who went to the sauna multiple times a week. But for people who *can't* exercise, going to the sauna might be a good first step toward heart health. And for those who do exercise, going to the sauna is an additional perk for the heart.

Using the same cohort of Finnish men studied over 30 years as well as a newer cohort of women, researchers led by Jari Laukkanen at the University of Eastern Finland found that going to the sauna regularly also reduces the risk of stroke for both sexes. His team is investigating whether the same holds true for other aging diseases, such as Alzheimer's. Meanwhile, others are investigating whether saunas, as well as traditional sweat-lodge ceremonies, can help people overcome addiction to drugs and alcohol, and whether they help troubled youth develop healthier coping habits.

A psychologist in New Mexico, Stephen Colmant, who has done some of the few published studies on sweat psychological therapy, told me that the idea is based on the fact that being in a hot space can be physically and psychically challenging. So challenging that a person needs to develop perseverance and serenity in the process— qualities that could help individuals overcome a multitude of other difficulties.

"At first, the heat is soothing and as the body begins to respond to the heat through sweating, the body's muscles experience a release of tension, promoting a deeper state of relaxation," Colmant and colleagues have written. "As the heat becomes more intense, the participant is challenged to keep the mind relaxed, requiring meditative attentiveness. As the experience moves from relaxation to endurance, it seems that participants are faced with a choice. One can either allow negative thoughts and feelings related to the heat to become the focus of their experience or one can focus on thoughts and feelings that help one to adapt, cope, and thrive when faced with adversity." In other words, the logic is: If you can learn to endure the heat, you can learn to cope with life's challenges and be more confident with your ability to do so.

It is certainly a challenge to spend time in a sauna—in many of my first Aufguss sessions I wondered whether I had the stamina to stay for the whole routine. Getting over the mental challenge brings a sense of deep accomplishment. Surviving a few Aufguss routines—just as a spectator—also brings a welcome physical exhaustion. I always have an incredible night's sleep after a sweat session.

After dozens more Aufguss World Championship performances—which featured stories about moonshine makers during Prohibition, Cruella de Vil and the 101 Dalmatians, and an F1 race car driver—Žákovský won the singles competition, becoming an Aufguss world champion.

"Getting into Aufguss was a coincidence," he tells me. "I was working as a lifeguard at a gym and the sauna's Aufguss master asked me if I was interested in doing the Aufguss infusion at some point. I said, 'No, I'm not going to play the fool by waving a towel around.'"

His coworker kept cajoling him, and Žákovský finally agreed to step into the sauna. "I got hooked. Looking at all the happy faces in the sauna during the Aufguss I thought, 'I want to do a job that makes people happy.'"

....

"I hate Aufguss theater," says Risto Elomaa, the head of the International Sauna Association, when we meet over a coffee on a snowy January day in his hometown of Helsinki, Finland. "Traditional Aufguss is great. But theater in a sauna—no."

Among Finns, he's not alone in his disdain: In the sauna theater World Championship's decade-plus history, I could find no evidence that there has ever been a participant competing under the Finnish flag. Nearby Norway has a strong Aufguss theater scene, and the World Championship has featured participants from countries as distant as Canada and Malaysia. But Finns react with a mixture of scorn or amusement when you tell them about the Aufguss. While sweating in Helsinki at the lovely neighborhood Arla sauna, I ended up in conversation with a woman next to me. When I told her about Aufguss theater, she began laughing so hard she choked. "They get dressed up and dance around with a towel in the sauna? Well, that's *awkward*."

For Elomaa, his disdain for Aufguss theater is about maintaining the dignity of the sauna experience. "In the sauna, you are sweating. You are throwing *löyly* [the Finnish term for steam that blasts off hot rocks when you pour water on them]. You are relaxing. You are forgetting everything—the whole outside world. And when you come out, you are a new person. That's how going to the sauna should be. People attending these Aufguss competitions—they are just going to the theater. They are not enjoying the sauna just for the sauna itself."

Finns are often stereotyped as quiet and reserved. Elomaa is neither. The outspoken retired chemical engineer does not mince his words. Eija Elomaa, his wife, teasingly bemoans the fact that she married "Mr. Sauna himself" after he embarks on an extensive rant about some software engineers who approached the sauna society recently about developing a sauna video game.

"This idea is completely stupid. You are not actually going to the sauna. You are sitting at your desk. You throw löyly virtually on the computer. On screen it shows the temperature and humidity. And you decide whether you can stand the heat or whether you have to leave the virtual sauna. I really hate this idea. I want people to go to a *real* sauna."

Finland certainly has a lot of real saunas: The country has more than 3 million of these sweat sanctuaries for a national population of only 5 million people. It is an understatement to say that Finns venerate the sauna. It's a place that, until the twentieth century, was incorporated into every aspect of Finnish life: Many were born in a sauna (it is warm and sterile); they smoked meat in the sauna; they rid clothes of fleas in the sauna; they died in the sauna. The country understandably feels an ownership of the sauna: It's arguably the only Finnish word that has made it into common international parlance.

The idea of relaxing in a heated room and pouring water on rocks to make it steamy was not, however, invented in Finland. There is archaeological evidence that steam baths existed everywhere from Pakistan to Mexico. The tradition could have traveled to Finland from the Americas across the Arctic or perhaps its presence there is thanks to Middle East travelers, or to the Romans, who loved steam baths, or to the Turks and their fabulous hammams. Or maybe, the pleasure found in sweating is so universal that humans around the world developed the practice independently. As Elomaa puts it, "To invent the sauna is not so special. There is an oven, there are some stones, there are some benches and that's it."

But Finland is responsible for popularizing the small private sauna in modern Western society. While other medieval Europeans were reluctant to remove clothing, fearing that nudity might make one more susceptible to plague, the Finns were still going naked into the sauna. For millennia, the sauna has been "a living culture in Finland," Elomaa says.

Which is the root of a certain understandable prickliness I've noted from Finns, that it's not specifically the *Finnish* sauna tradition that's gaining traction in international wellness worlds: it's the German Aufguss ceremony, which typically takes place in a Finnish sauna. Case in point: At a recent Global Wellness Summit, run by the Miami-based Global Wellness Institute, the Aufguss was named #1 of the year's top eight wellness trends, followed by other activities such as "silence" (as in silent meditation and the like) and "art & creativity wellness" (a.k.a. adult coloring books). If you haven't already, you will probably have an opportunity to try out an Aufguss at a spa near you in the next decade.

This bustling Aufguss industry owes its existence in part to the Finns, who, at the turn of the twentieth century, helped reintroduce sweating for fun to the rest of Europe and set the stage for Germany's invention of the Aufguss. I have yet to find any serious historian who has unequivocally tracked the origin of the Aufguss. But if you ask around, someone inevitably points to the 1936 Winter Olympics in the Bavarian town of Garmisch-Partenkirchen, during the height of Nazi Germany. As the story goes, German athletes did not win as many medals as Finnish athletes, and Hitler was pissed off. Consequently, a subordinate was dispatched to find out why and came back with tales of the Finnish athletes relaxing in saunas after their workouts. And thus the Finnish löyly tradition—a practice of going into a dry, very hot sauna to create steam by throwing water on the hot rocks—was reintroduced to Teutonic attention. But here's the thing: Finland *didn't* win more medals at the Garmisch-Partenkirchen Olympics. Finland and Germany had the same overall medal count (six apiece), and German athletes won more gold and silver than their Finnish counterparts. The big winner was Norway, with 15 medals—of which seven were gold. Like the Finns, Norwegians enjoy a good sweat, so perhaps someone conflated the two nations in the creation of this probable myth. However, as with many myths, there are threads of truth to be found.

Certainly Finnish athletes built a sauna at the Olympic Games in Nazi Germany, as they commonly did at all major sporting events. But the mystique of the sauna had likely already spread beyond Finnish borders in the 1920s. Tuomo Särkikoski, who has written a comprehensive history of the Finnish sauna, thinks that Finnish runners who dominated Olympic Games in the 1920s helped to popularize the sauna abroad. Särkikoski points to runners such as Paavo "Flying Finn" Nurmi, who won 12 Olympic medals over the course of his career and set 22 world records. Although Nurmi often avoided the public eye, purportedly claiming that "worldly fame and reputation are worth less than a rotten lingonberry," he was not shy about his belief that sauna sessions helped him beat his competitors.

During the 1936 Olympics in Nazi Germany, Finnish athletes recuperated in saunas, which Third Reich propagandist Leni Riefenstahl caught on film. "Given her mandate to make the Aryan figure popular, she used the image of the [very buff] Finnish athletes in the sauna as an expression of Aryan superiority," Särkikoski says. The collaborations between Finnish sauna afficionados and the Third Reich didn't end there.

Särkikoski found archival evidence that Heinrich Himmler, one of Hitler's closest advisers, had a special interest in saunas, particularly in whether sweating sessions could improve fertility, an Aryan eugenic goal. Himmler also corresponded with Finnish sauna experts about whether mobile saunas might be useful for soldiers on the move.

After World War II, most Europeans didn't have the money to go to a fancy spa on the weekend, particularly not in Germany where the war had bankrupted the nation. But by the 1980s, in German-speaking areas, relaxing in a sauna was taking off as a wellness idea on both sides of the Berlin Wall. A German friend of mine who grew up in former communist East Berlin is nostalgic about the mobile sauna box that her family would set up on Saturday nights in the 1980s in their apartment's living room. There was room for only one person to sit inside the sauna box from the neck downward; the sweating person's head remained outside thanks to a rubber turtleneck-like device. The family of three would take turns in their living-room sauna on weekends, with two family members cooling off on the pull-out couch while one person sweated from the neck down inside.

By the 1990s, sauna culture was widespread, and the Aufguss, with its curious towel routines that distributed steam around the room, began appearing at spas around Germany, Austria, and German-speaking Südtirol in the Italian Alps.

. . . .

Another sauna spin-off industry gaining enormous business traction is the infrared sauna, a market worth $75 million in 2017. Except infrared saunas are not technically saunas. Calling them so will earn you a lecture from the president of the International Sauna Association. "Don't call them infrared saunas. These are not technically real saunas," Elomaa says. "Please call them *infrared sweat cabins.*"

Elomaa sounds like he's splitting hairs but his position is legit. In 1999, the international sauna community gathered in Stuttgart from places as far away as Japan to agree upon a strict definition of what you can call a sauna, and as a result of the process, infrared technology got cut out of the club.

The international sauna Illuminati had many motivations for coming up with a strict definition of a sauna, Elomaa says, including a desire to solve a salacious reputation problem.

"[In 1999,] Germany was full of saunas that were actually bordellos," Elomaa explains. "You would go to Hamburg and see a sign for sauna. But the maximum temperature inside would be barely enough to break a sweat." To dissuade the sex industry from moonlighting on their territory, the International Sauna Congress decreed that a sauna needed to have, ideally, a temperature of between 167°F and

175°F at 1 yard above the bench (or 75°C and 80°C 1 meter above the bench in Europe's preferred metric units).

Decreeing a minimum temperature requirement helps with hygiene too. "You should have, on the floor level, a temperature of at least 135°F because otherwise there can be bacteria and fungi growing on the floor, especially if the sauna is humid," Elomaa explains. Above the minimum temperature threshold, everyone can safely walk around barefoot in the sauna and not worry that they'll leave with a foot fungus infection, even if someone else arrived with one. (Researchers at the Finnish Sauna Society in Helsinki figured out this temperature cutoff back in the 1970s when the organization had its own laboratories and research collaborators at local hospitals. In typical Cold War fashion, these researchers also evaluated whether radioactivity from nuclear fallout could be excreted in sauna sweat. The results were not promising, which is not surprising given that the kidneys deal with chemical detoxification, and radionuclides are chemicals.)

Though many infrared enterprises offer temperatures below the official sauna minimum, it is a different component of the 1999 sauna decree that negated their membership in the club: *stones*. "A sauna is a room with wooden walls. There has to be an oven, and here's the key thing, there has to be stones. And you must have the possibility to throw water on the stones," Elomaa explains.

Infrared cabins have no stones. If you threw water at the heat source, an electrical fire might ensue, or a short-circuit at the very least. What gets passed off in the United States as an "infrared sauna" is effectively just an enclosed space warmed directly with a space heater—you know, the same style of plug-and-play heater used to warm a cold office or to grill meat on a spit. To be as blunt as Elomaa, "It's like buying a Ford versus a Ferrari."

....

If infrared heat is considered lowbrow by some sweat aficionados, the high end of the sweat bathing experience is occupied by old-school smoke saunas, particularly those found at the Finnish Sauna Society's headquarters. Sweat nerds speak in hushed tones about Saunaseura, the expansive smoke sauna complex nestled in a forested area on the edge of the Baltic Sea on the outskirts of Helsinki. It's a place where royalty, heads of state, and diplomats get their sweat on, along with average Joes, *if* you happen to be a Finnish Sauna Society member or a guest.

I wanted to visit the Saunaseura but sweatiquette prevented this. In Finnish public saunas, you typically don't sweat naked in mixed company (unless you trek down an industrial peninsula in Helsinki's harbor and sweat at the delightfully anarchist, free public sauna called Sompasauna). Saunaseura has separate women's days and men's days so that nobody has to wear anything constricting or be scandalized by naked bodies of the opposite sex—which is in stark contrast to Germany and the Netherlands, where it is so commonplace for everyone to be naked* in the sauna that the formality (or, actually, informality) can surprise unsuspecting tourists.

But herein lay a conundrum. At the Saunaseura, only members of the Finnish Sauna Society can bring a guest to their sauna complex, and the only member I know—Elomaa—is a man. So Elomaa recruited Eija Elomaa, his wife, to escort me inside, a task she did with grace despite the fact that she was busy organizing a quilt exhibit. When I ask if it is annoying to babysit a journalist for the afternoon, she says, with a wry grin, "You're not the first one."

---

* Most regular sauna goers would agree that sweating naked is inherently more pleasurable than sweating in a bathing suit—but of course that requires feeling comfortable being naked in public spaces. Saunas that have specified days for either men or women may make some segments of the population more comfortable with public nudity, but the practice is effectively exclusionary to people with non-binary bodies or those with bodies in transition. Enforced nudity for all people may get around the inequity of binary categorization, but it forces people who may not wish to bare all to do so.

We arrive at the Finnish Sauna Society in the early afternoon. In January, in Finland, early afternoon means that the sun is already starting to set, casting long shadows across the wintry landscape whenever its light manages to penetrate the cloudy sky. There is nothing around the isolated sauna complex but snow-laden birch and evergreens leading to the Baltic coast. A long dock stretches into the water past frozen seagrass. Wood-fire smoke and the smell of the sea fill the air, while the temperature hovers at a chilly 10 degrees below freezing. "Will you swim in the Baltic today?" I ask her, as we walk from the parking lot toward the entrance. "Of course," comes her answer. "And you?"

"Of course," I parrot, gulping inwardly.

Like many traditional saunas, smoke saunas are heated with fire-wood. But instead of just heating the sauna's rocks using a wood-burning stove and venting the smoke out through a chimney, the smoke is channeled through the interior of the sauna for the 5-plus hours it takes to get rocks hot enough to sustain temperatures of about 200°F. After the rocks are heated, the interior is hosed down so that sauna goers don't get covered in black soot. The result is a dark, smoke-seasoned interior with a luxurious woody scent and an emotional warmth that goes far beyond the searing heat. Elomaa told me that part of the appeal of the Saunaseura was its special sauna rocks, a gray-green volcanic mineral mix called peridotite exclusively sourced from an old mine in the tiny Finnish town of Orimattila.

Sauna rocks must have excellent heat capacity—the ability to withstand high temperatures without immediately cracking, as well as the ability to hold that heat so that the sauna stays warm. Which is why volcanic rock is the go-to source: The temperature of a sauna pales in comparison to the scorching blaze of Earth's interior.

The most common sauna rock, called diabase, won't win any beauty contests. You wouldn't be tempted to collect the gray, rough stone unless you were a sauna owner looking for a free restock. (Though the builders of Stonehenge liked it enough to use it for that

masterpiece.) But let's face it, in the low light of a sauna, most people don't care what the hot rocks look like, and diabase is sturdy,
affordable, and, importantly, it's not very porous. That means water
thrown at these rocks vaporizes off as steam instead of disappearing into the stone's interior.

Stones, pretty or plain, will eventually crack and need
replacement—every few months in a heavily used spa sauna and
every few years for a private sauna. Cracked sauna rocks result in
crumbly dust that can tag along with the water vaporizing into the
air during a löyly or Aufguss, which means the dust could eventually
settle in your lungs—not what anyone wants from a sauna visit.

In the changing room, Eija nods appreciatively at my birch
branch–emblazoned felted sauna hat. I feel my sauna credibility
rise ever so slightly in her eyes. *At least she is spending her afternoon with a journalist who has her own sauna hat.*

As is traditional in Finnish saunas, we first go to the shower room
and drench ourselves. To one side of the showers, Eija shows me an
alcove where you can be scrubbed silly for a small fee—it's amazing
the amount of dead skin that comes off after sweating heavily for a
few hours. After such a scrub—which you also find in hammams,
jjimjilbangs, and many sweat institutions worldwide—your whole
body is as smooth as a baby's.

On the other side of the showers is a passage to the relaxation and
cafeteria restaurant area, where you can recover from your sweat
with some food around a fire and a huge glass window that surveys
the sauna's grounds and waterfront.

"Shall we go in?" Eija says and pulls open a small wooden door.
Inside it is so dark from the soot seasoning that I can barely see the
sauna's rock oven on my left and the six women inside, sitting on a
platform a few wooden steps up from the entrance. I inhale deeply
and feel the warm, richly pungent air fill my lungs. It smells delightful, the smoke a pleasant afterthought. As my eyes adjust to the
minimal light, which is coming from a small window and several

vents, I notice a wooden bucket and ladle with a birch branch soaking inside. The women briefly pause their conversation to acknowledge our arrival with a nod, and then continue an animated debate. Eija leans over and translates in a low voice. "They're annoyed about some construction on the subway." Hearing the English, the women switch languages effortlessly to include me in the conversation.

In most spas, the sauna is sacrosanct. You go there to relax, for solitude. If you go with a friend to a sauna in Germany and you get to chatting, people glare, then shush you. Nobody successfully starts conversations with strangers in a German sauna. Yet in Finland, there we all were, naked, getting to know one another. I had once read that if you need to get a Finn talking, put them in a sauna.

"How about some löyly?" Eija asks, looking around the room. Everyone nods. She reaches for the ladle in the bucket and scoops some water onto the hot rocks. Immediately a burst of steam pulses up and off the oven. Amid the sauna's woody, smoky scent, I detect the aroma of birch in the hot wind, a gift from the birch branch, which had infused the bucket's water with its odor. Unlike in the German Aufguss where essential oils are used widely, the only added aroma in a Finnish sauna (aside from smoke) comes from whatever has leached out of the birch or pine branches.

Eija pours another scoop of water on the rocks and a second pulse of steam envelops us all. The room falls silent as everyone inhales the humid air. Beads of sweat on my back turn into a full-blown flood. I ask if I could use the *vihta*, the birch branch soaking in the bucket. Slowly I begin to whip myself, which results in some additional hot wind on my skin followed by the sting of impact where the wet leaves hit my body. It feels delicious, like a self-inflicted masochistic massage. We all sit in silence for a while, enjoying the searing heat and the natural scents. After a few minutes, or maybe 10, I begin to feel like I am reaching my maximum. Eija turns her head and peers at me with expert eyes: "You are ready to leave? How do you feel?"

"Incredible. But yes, a little light-headed."

"Shall we go for a dip?"

I had forgotten about the part where I was supposed to jump into the Baltic Sea in the dead of winter.

"Okay!" I say, more enthusiastically than I feel. Bidding adieu to the other women in the sauna (I wonder when they will hit *their* max?), we head out onto the sauna's terrace. The freezing air feels unexpectedly refreshing. I wrap myself in a sauna towel and stand with a handful of other women surveying the water. On a digital display, I see that the air temperature is now –7°C (19°F) and the water temperature is 2°C (36°F).

"It's now or never," I say, and follow Eija at a brisk pace down a path from the terrace to the dock.

If I am going to do this, I don't want to lose my mojo with a slow saunter. At the end of the dock we ditch our towels and flip-flops. I grab the ladder railing and begin to lower myself into the water. Amazingly the water feels warmer than the air around us. Because it *is* warmer than the air around us. I spring off the ladder and dive head first into the sea.

When I pop up a second or so later, the first thing I see is Eija's concerned face peering over the dock. Her eyes wide, she says "Oh!" and exhales loudly as I emerge from the water. "Are you okay? You know, we don't usually put our heads under water! It's not good for the—," and with this she taps her cranium.

"It felt absolutely great," I answer, climbing back up the stairs, and I wholeheartedly mean it. She lowers herself in and carefully treads water, keeping her head above the surface before we both shuffle back up the path and inside.

By the time we get back to the terrace we are both ice cold. Eija suggests we warm up by the fire in the restaurant with some steaming salmon soup. While waiting, I have a look at letters from Britain's Prince Philip and a former US ambassador to Finland posted on the wall. It reminds me of what Elomaa had told me about Finnish sauna diplomacy.

"In the sauna you have no weapons. It is a safe place. The sauna is also a telephone-free area." With little risk of cyber surveillance, people can talk freely, he explained. But the main premise of sauna diplomacy is to get your adversary into the sauna and push them to sweat until they are more amenable to negotiation. "Going to the sauna is part of Finnish hospitality; it's the social part of any diplomatic visit," Elomaa had told me.

"Not everyone accepts the invitation, but if you have a visitor from Russia, they normally like the sauna." Finland has long had a loaded relationship with its big eastern neighbor. Over the past centuries, Russia has had a repeated tendency to occupy Finland. And even though Finland has been independent since 1917, Russia has often tried strong-arming its neighbor, particularly during the Cold War. When I asked Elomaa whether sauna diplomacy had ever actually led to any specific diplomatic achievements, he immediately jumped to the late 1970s, at the height of the Cold War.

"Dmitry Ustinov was the most important general of the Soviet Union army [and minister of defense]. He came to meet with [then Finnish president Urho] Kekkonen because the Soviet Union wanted to initiate some joint military maneuvers with Finland," Elomaa said. "And Finland was absolutely against it." That's because the country wanted to remain on good terms with the rest of Western Europe, which would certainly feel threatened. The Russians were putting a lot of pressure on Kekkonen to comply, Elomaa said, and they continued discussions late into the evening in Kekkonen's summerhouse sauna.

"I don't know what was said in the sauna," Elomaa said, "but the next night Ustinov left. Finland did not agree to maneuvers and the Russians never came back to this question."

Is this story true? Maybe. Or maybe not.

According to Finnish historian Hannu Rautkallio, an earlier Soviet leader, Nikita Khrushchev, did join the Finnish political elite in the sauna a few decades earlier, in the 1950s. When Khrushchev returned home, he was criticized for having cavorted naked in a sauna with the bourgeoisie. To save face with party members, Khrushchev assured everyone that he had not, technically, been naked in the sauna with a capitalist. He had worn shorts.

. . . .

After Eija and I finish some salmon soup, she leaves me alone at the Saunaseura and returns home to get some work done. "Enjoy yourself," she says with a smile that somehow includes an unspoken suggestion to be judicious with the whole dunking my head under water routine.

After we say our goodbyes, I go back to the sauna area to get my sweat on again. This time, I am determined to try pouring the water on the rocks myself, to get some löyly. Two women in their thirties are chatting in Finnish when I enter. After a little while, I ask if they would like a löyly. Eija has told me that you can't make the sauna hotter with steam if everyone doesn't feel comfortable with the idea. In practice, most people agree that it's okay or they use the opportunity to exit for a freezing dunk.

Permission granted, I scoop up some water and lean over the rocks.

"Stop! Stand back!" exclaims a voice in English.

I turn my head to see one of the two women looking at me with alarm. "You have to stretch out your arm and turn your head away so you don't get burned by the steam pulse," she says. "And pour the water slowly. Very slowly."

"It's my first time doing the löyly," I admit.

"I guessed that," she replies, not unkindly.

I follow her directions, and even so the warm air hits me with a furious blast. I would have scalded my face if she hadn't intervened. Instead I smile gratefully, sit back, and inhale the glorious, swirling steam.

6

SWEATPRINTS

In 2016, West Yorkshire Police in the north of England were called to investigate a case of breaking and entering at a woman's home. Police were able to lift the perpetrator's fingerprint from her window frame, which they linked to a man who had been stalking her. Investigators also sent the print to Simona Francese, a chemist at Sheffield Hallam University who specializes in analyzing trace chemicals left behind in fingerprints, which are, at their essence, just sweat marks.

When we leave fingerprints behind, they are an impression of our fingers or thumbs produced with our own translucent biological ink: that chemically complex cocktail of molecules released in a fluid called sweat. When Francese and her team analyzed the ridges of the crime scene fingerprint, they found chemical traces of cocaine, suggesting that the man who broke in was high at the time. They also found something more unusual: a molecule called cocaethylene. When a person snorts cocaine and drinks alcohol at the same time, the two intoxicants circulate in that person's blood simultaneously. When the blood reaches the liver, the organ tries to break down both drugs. During this metabolism, the liver creates a hybrid molecule called cocaethylene from partially broken-down alcohol and cocaine. That marker traverses back into the bloodstream and can eventually make a cameo appearance in sweat, as it did in this harasser's fingerprint.

"Alcohol intensifies the effects of cocaine," Francese says, "which gives you an idea about the state of mind of the individual whilst committing the crime."

Back at the police station, the man tested positive for cocaine, Francese learned later, and eventually admitted to drinking alcohol as well—confirming the information she had found in his fingerprint.

Anyone who has indulged in a garlicky dish or a hard night of drinking knows that the stuff you binge on can seep out in your sweat. Sometimes it has an aroma; more rarely, it has a color. But what about everything else? The hundreds of odorless, colorless chemicals percolating out in your sweat that reveal the drugs you are taking, both pharmaceutical and otherwise, as well as other hints about your identity, health, and well-being?

Private truths about your lifestyle are left behind—albeit in trace quantities—in your fingerprints. And thanks to sophisticated analytical technology, researchers are now able to deduce these secrets from the smallest traces of sweat, even those found in a fingerprint.

Law enforcement has been peering at fingerprints since the late 1880s. Francis Galton, a cousin of Charles Darwin, popularized the idea that everybody had unique patterns of whorls and swirls on their fingers that could be used to identify them. Over the past century, police have focused primarily on finding ways to *visualize* these sweaty marks so that fingerprints at a crime scene can be compared with those of potential suspects.

Many forensic techniques devised to make colorless fingerprints visible to the naked eye rely on their sweaty chemistry. Case in point: the classic trick of using ninhydrin dye to stain fingerprints a vibrant purple-pink color. Sweat coming out of eccrine pores contains trace amounts of proteins and amino acids; ninhydrin reacts with these components to stain the fingerprints bright magenta.

Likewise, when forensic teams use silver nitrate solutions to turn a fingerprint from translucent to black, they rely on the fact that salty components in sweat, namely chloride ions, will react with silver nitrate to form a very noticeable black compound.

Law enforcement officials want a beautifully resolved fingerprint so they can match its pattern with those of a potential suspect or with one in a criminal fingerprint database. But what if crime scene fingerprints have no matches whatsoever? The fingerprints may be perfectly visible, but their sharp looks aren't enough to move the case forward if the person in question has never been detained or had their prints entered in a database. This issue inspired forensic researchers to wonder whether additional information could be gathered from the chemicals left behind in the fingerprints themselves. In 1969, the United Kingdom's Atomic Energy Authority published a report that examined the fingerprints of 500 people to see if useful information could be garnered from the most abundant ingredient of eccrine sweat: its salt.

The report focused on concentrations of chloride ions found in the fingerprint sweat. The idea was based in some solid logic: Cystic fibrosis can be diagnosed by chloride levels in sweat because people with the disorder produce perspiration with higher-than-average levels of salt. The scientists searched for additional correlations between chloride levels in fingerprints and characteristics of the person who left the marks behind, such as the person's age, sex, and employment. The research didn't lead to any new forensic techniques, in part because researchers focused on the wrong fingerprint chemical. Salt levels can be extremely variable among healthy individuals—some people have such salty sweat that they could be misdiagnosed as having cystic fibrosis.

The forensic scientists of 50 years ago probably focused on salt because it was a component of sweat that they could easily measure. At the time, analytical technology was simply not sophisticated enough to accurately measure trace amounts of other fingerprint chemicals like proteins or hormones. The trace levels of these molecules in fingerprints were beyond the capabilities of their machinery.

But by the mid-2000s, all that began to change. Laboratory instruments were becoming so advanced that forensic scientists could hope to see—just from the fingerprint's chemical makeup— whether the person who left the mark had recently indulged in drugs, such as cocaine, or more benign intoxicants, such as caffeine.

Since then, law enforcement agencies, including the United Kingdom's Home Office, have funded researchers to study all the chemicals in fingerprints, to investigate whether characteristics such as sex and age could be identified in the marks. Law enforcement agencies have also been giving scientists like Francese access to fingerprints from ongoing and cold cases to see what additional clues can be extracted. Forensic fingerprint chemical analysis is still in its infancy, but if the techniques reach their full potential, fingerprint chemical analysis may have as much impact on criminal cases as DNA sequencing.

Which is why I found myself traveling to Sheffield, England, to see what Francese might find in my own sweaty fingerprints.

....

Famous for its steel and its dismal architecture, Sheffield has been belittled by critics for centuries. The French duke Rochefoucauld, visiting in 1785, wrote that "Sheffield could never pass for a fine town" thanks to its "shapeless huts and outlandish factory-buildings." Nearly a hundred year later, writer Walter White quipped, "What a beautiful place Sheffield would be, if Sheffield were not there!" Even George Orwell threw shade at Sheffield. "It could justly claim to be called the ugliest town in the Old World," Orwell wrote. "And the stench! If at rare moments you stop smelling sulphur it is because you have begun smelling gas." Those cruel words were written in 1937, before the Germans heavily bombed the industrial town during World War II.

Just outside the Sheffield train station, I trudge up a winding path through a neglected park to get to my B&B, and I press "play" on Cabaret Voltaire, one of Sheffield's celebrated electronic bands. (Sheffield was the cauldron for "synth pop," but what might have been the beginning of an urbane musical tradition went terribly awry when the city spawned the cheesy hair-metal band Def Leppard.) Among blooming poppies, there's a skinny Neo-Gothic monument to commemorate Sheffield's cholera victims. *This town just can't get a break*, I think. Even the monument's pinnacle was struck by lightning in 1990 and needed replacement.

Amid an otherwise generic cityscape skyline, there are four giant steel buildings shaped like curling stones—the ones cajoled along ice by broom-wielding Olympic athletes. These curious buildings are the student union of Sheffield Hallam University, where I head the next morning to meet Simona Francese.

Sporting jeans, high heels, a red lab coat, and a melodic Italian accent, Francese has spent the past decade working on finding ways to get as much chemical information as possible from a fingerprint. "I was always interested in forensics," she tells me as we traverse a confusing maze of staircases and buildings en route to the lab, "but at the time I started university in Italy, there was no forensic sciences degree there." Instead, Francese studied chemistry and began using mass spectrometry to research medical questions, such as how pharmaceutical drugs penetrate skin.

Initially she used pig skin to study drug absorption. When it was time to move on to human skin, Francese applied to a local tissue bank for permission to get a human sample. And then she waited. "Finally a box showed up in my office," she says. "I was expecting the hospital would send me a few pieces of human skin from random places in the body. But when I opened up the box, an entire breast was just sitting there.

"I stared at it for quite some time before calling my colleagues to start the experiments," she says. Shortly thereafter, Francese decided to focus on what comes out from skin on fingerprints instead of studying how drugs sink in.

"Fingerprints have always really fascinated me. They are so beautiful. And to people who are trained in chemistry, it's obvious that fingerprints aren't just inanimate objects. There is organic and inorganic matter there to be discovered."

In my case, Francese would track my penchant for caffeine by watching it appear in my fingerprints after I imbibed a coffee. I had originally been drawn to the idea of making NikNaks Spicy Tomato red flavoring emerge in my sweat, but that plan was thwarted by the fact that the flavor has long been discontinued. Monitoring how caffeine percolates out in my fingerprints sounded like a fabulous alternative. That is, until it became clear that I would have to navigate the route to Sheffield Hallam University without the guiding benefits of my morning brew: Francese wanted to start the experiment with a clean, caffeine-free journalist, so we would have some virgin prints as reference.

Francese takes me to at a bright laboratory full of windows, humming equipment, and computers. There she introduces me to Jillian Newton, a staff scientist at the university's mass spectrometry facility who will be running the machines used to analyze my fingerprint chemistry. I am not the first journalist she has worked with on a body-hacking project, and mine is relatively tame: Her most memorable project involved studying the stinky emissions of rotting chickens. The project was the brainchild of a BBC producer who asked Newton to calculate the amount of methane wafting off a decomposing chicken, as part of a project to assess the role of rotting food in greenhouse gas emissions. And yes, you can find a time-lapse video of the decomposition online. As luck would have it, the rotting odor wafted over to the university's janitorial staff offices nearby. "These are really not people you want as enemies," she said. "So sometimes my job involves quite a bit of diplomacy."

Today, Newton has been pondering a different conundrum. She has been tasked with freezing mosquitoes and then cutting them in half. The idea is to study the biological molecules present across the entire thin slice of insect. But as soon as the knife hits, "the fragile, crinkly bodies just shatter," Newton explains. Her next plan is to embed the mosquitoes in a malleable material that can hold the insects stable as they are cut. Once she has a nice slice, Newton will scan the surface of the insect's body using the same machine, a mass spectrometer, that she'll use to measure the chemicals present across the surface of my fingerprint.

Francese reappears with a lab coat for me. After I put it on, she places a small metal wafer on the table in front of me. "Press here," she says. "Any finger. Not too hard. But hard enough to leave something behind." Newton and Francese stare at me expectantly while I focus on not screwing up the incredibly simple task of leaving a fingerprint. I choose my index finger and leave a barely visible trace of finger sweat on the wafer's surface.

Newton puts the wafer in the machine, and a laser scans over the fingerprint. The laser's job is to gently blast the molecules present in the fingerprint into the air, without entirely destroying them, so that they can be dispatched into another part of the machine for analysis. She has additionally sprayed my fingerprint with a polymer matrix coating that will help to dispatch the chemicals found in my sweat along the machine's flight tube.

Once the molecules are airborne, you can measure their mass (which is why the machine is called a mass spectrometer). The net result is a complete list of all masses of molecules that were kicked off my fingerprint. So for example, when I have caffeine in my system, the researchers expect to see an entry at 195, which corresponds to the combined mass of 8 carbon atoms, 10 hydrogen atoms, 4 nitrogen atoms, and 2 oxygen atoms present in one molecule of caffeine, or $C_8H_{10}N_4O_2$, plus a component from the polymer matrix.

As Newton takes data on my pre-caffeine fingerprint, Francese gives me some good news. "Let's go get you some coffee."

At the café a few floors down, Francese tells me about her collaborations with law enforcement in the United Kingdom and the Netherlands while I sip a flat white. Fingerprint chemical analysis is still being refined by academic scientists, she says, but in the meantime, "police here and abroad are willing to give us some samples from ongoing investigations to see what we can do."

In addition to working with police, Francese has also collaborated with a drug rehabilitation clinic to test the fingerprints of individuals in recovery from opioid use. "You'd expect to see methadone in the fingerprints," Francese says, "but we also found cocaine." Namely a set of three molecules with masses of 304, 290, and 200, corresponding to cocaine and two chemicals, ecgonine methyl ester and benzoylecgonine, that are produced by the body when it breaks down the drug. Francese's results suggest that some of the clinic's patients were supplementing with cocaine.

Francese looks at her watch and my empty coffee cup. It has been about an hour since I started drinking the brew. "Let's go see if we can find some caffeine in your fingermarks." (People in Britain sometimes refer to fingerprints as fingermarks.)

Back in the lab, I am feeling a lot more confident about my fingerprint-depositing capabilities. After I get the job done for a second time, Newton sprays the wafer holding my caffeinated fingerprint with some protective polymer and slips it into the mass spectrometer. And then she turns on the laser and directs its beam over my fingerprint.

On the computer screen in front of us, we watch as the mass spectrometer tabulates the presence of hundreds of molecules in my prints. Newton scans through the data, as I sit watching, overwhelmed by how many molecules the machine is detecting. Intellectually, I know that hundreds of chemicals come out in my sweat. But here, right in front of me, in the smallest traces of my sweat, in my fingerprint, are piles of molecules, my molecules, appearing on-screen. It is unnerving to see how leaky my body is.

At the entry for caffeine, a huge peak appeared. "There it is," Newton says.

"Oh my God. I should never commit a crime."

Francese chuckles. "Well, certainly not if caffeine would incriminate you."

....

Forensic researchers like Francese aren't just interested in finding clues about drug consumption—legal or otherwise—in fingerprints. They are also keen to learn about people's lifestyle habits. For example, are you a strict vegan or a meat eater? People switching between these diets often report that their body odor changes. Because odor is just a collection of sweat chemicals we can smell, there are probably many more chemicals emerging in your sweat that reveal these food preferences that don't have an odor, but which could be detected by sensitive analytical equipment.

Or what about your preference for birth control? Francese and her colleagues have found that they can identify a person's condom brand preferences by analyzing the trace levels of lubricant left on that person's fingers, which then get mixed in with sweat and deposited in a fingerprint. Francese told me that with the advent of rape kits and DNA testing, sexual predators have taken to wearing condoms so that they don't leave behind any semen, and thus any DNA. Detecting condom lubricant in a fingerprint isn't a smoking gun, but it certainly supports a rape survivor's account, and it might narrow down a list of suspects. As would trace amounts of other unusual chemicals used by criminals—skin moisturizers, sunscreen, bug spray, or other personal care products that people spread over their fingers over the course of a day.

Then there's the stuff coming out in your sweat that you don't have control of, chemicals that may reveal your sex or health status. Several groups have found that men and women release different levels of proteins and peptides in their sweat. These bulky molecules are made by our immune system for good reason: They are natural antimicrobials, released from the sweat glands onto skin as a first line of defense against tiny pathogens. Both men and women release them, but thanks to hormonal differences, the relative abundance of these compounds in sweat can point to your biological sex. Francese's technique for distinguishing biological sex from fingerprints is accurate about 85% of the time. "This is pretty good. But in forensic sciences you need to be much more accurate than this. You have to be if you want to take a technique to court." So Francese is currently working to assemble a collection of about 200 fingerprints from a variety of donors to improve the accuracy of the sex-identifying technique.

Chemists at the State University of New York, Albany, are working on the same problem. Their preliminary research suggests that men and women might be distinguished by the relative amounts of amino acids coming out in their sweat. Women have, on average, about twice the levels of amino acids than those of men.

"I definitely want to nail down the sex thing," Francese says. "There's a silly idea that a majority of homicides are committed by men. First of all, it's not true, and second of all, you can't just dismiss the fact that a woman might have committed such a crime. Imagine the scenario that you can actually say, this person is a female, and she's from such and such ethnic background, and she is under these medications, and age 50 or thereabouts." Francese is also looking for disease markers in fingerprints that suggest a person is sick. "You can imagine someone having a very bad day," Francese says. "Because you could say, 'Well, I found out that you committed a crime and by the way, you also have cancer.' "

This fingerprint technique could be incredibly useful for police

investigators or for someone who might want to skip weeks of waiting for a cancer diagnosis. But what about privacy? Or the threat of
this sort of technology being used by a totalitarian regime? Or by
HR departments keen to avoid hiring people with otherwise invisible illnesses? "I understand these concerns," says Francese. "But I
believe that you can't really stop progress just because of the potential misuse of knowledge. I am not saying that there won't be a case
where somebody wakes up in the morning and uses [the technique]
in the wrong way. But I don't think that's enough of a good reason
to stop progress. For me, knowledge is the most important thing—
together with guided governmental use of that knowledge."

Will democratic governments respond fast enough to create laws
to prevent misuse of this new fingerprint analysis technology? Consider the case of surreptitious DNA collection. For years, police in
the United States and abroad have been collecting biological traces
of suspected criminals without first getting a warrant from a judge,
a hoop they'd have to jump through before legally setting up a phone
tap. Collecting conversations is arguably a lot less intrusive than
collecting DNA.

All a detective needs to do is watch a person of interest smoke a
cigarette, drink a coffee, or eat lunch. After the individual discards
the cigarette butt, beverage cup, or cutlery, the detective can swoop
in, collect the trash, and scrape off that person's DNA for analysis.
Advocates of the practice argue that when someone has knowingly
abandoned trash containing their biological residues, it's fair game
for surreptitious forensic collection. If collecting and analyzing a
person's DNA without a warrant is fair game, it's hard to see how
collecting and analyzing someone's sweat chemicals by way of their
fingerprints would be viewed any differently.

But not everyone agrees, and privacy advocates are on the DNA
case: Surreptitious sampling will likely see its day in the country's
highest court. Even news websites specifically for law enforcement are cautioning their readers to start seeking warrants for

DNA collection because, as this PoliceOne.com headline put it, "Surreptitious DNA sampling is knocking on the Supreme Court's door. Know the issues so your investigative practices can shape the debate and not result in unintended consequences in the form of a SCOTUS ruling."

In the absence of rulings from the highest courts, detectives are left to find their own way in the gray zone between good detective work and privacy invasion. Whatever is decided for DNA collection and analysis will set a precedent for chemical fingerprint analysis—hopefully before forensic analysis of fingerprints becomes widespread.

Meanwhile, it may be many years before the kind of techniques Francese is working on end up in a standard forensic kit. Francese believes a turning point will be when forensic fingerprint evidence gets its first day in court. "Because it's a new technology, of course [opposing legal teams] will try and discredit it because it has not been used before. My plan for now is to provide as much service to law enforcement [as possible]."

Francese opens up a file containing my caffeine-rich print. "Let's have a last look at your data and see what else is there," she says.

"Ooooh, what's that?" Newton exclaims.

Amid the hundreds of chemicals making cameos in my fingerprints, there is one chemical with a mass of 398 that was present in extremely abundant quantities.

"Huh. Now I'm curious," Francese says as she zooms in on the data for the chemical with a mass of 398. "I've never seen anything so abundant [in fingerprints] with that mass." She makes a humming-thinking noise. "Maybe you produce an unusual skin [molecule] that is mixing with your sweat. Or maybe it's the skin cream you use. I did see you touching your face before you made this mark." I feel a tad sheepish.

Francese directs the software to only show places on my fingerprint scan where this chemical could be found. It was everywhere. Every ridge held enough quantities of this chemical that you could see the outline of my fingerprint only using this weird, fatty molecule as the reference.

"So you're telling me that this may be a fingerprint molecule that only I produce?"

"Maybe it's a biomarker for psychopaths," Newton says.

"Oh great," I laugh. "My secret is out."

"Well," Francese adds, "just don't commit any crimes, because rare sweat molecules like this could incriminate you."

"You're really throwing a damper on my future plans."

"Good," Francese says. "Let me show you one last thing."

On the computer screen Francese opens a photo. It is a mess of overlapping fingerprints, as if a bunch of people had touched the same area. The fingerprint traces are so overlapping that you couldn't really distinguish one mark from another. And you could definitely not resolve any one, single fingerprint to do a database search.

"Imagine the situation where there was someone being assaulted by someone else. And say there was a fingerprint of the victim on a surface, and superimposed [was] the fingerprint of the assailant. At this point police would be stuck because they couldn't separate the two ridge patterns. But," and she pauses, "we can."

That's because, like me, everyone produces unique molecules in their sweat, which they leave behind in their fingerprints. When you look at the messy surface of these overlapping fingerprints, you can ask the visualization software to show only the presence of a unique chemical, a chemical that only emerges in the sweat of one of the individuals who contributed to the messy overlapping fingerprints. When you filter out all the other fingerprint chemicals, then the mess disappears—except for the fingerprint of the person producing that one chemical. This is how you can separate overlapping fingerprints digitally, which could then be fed into a finger-print database.

"You say to the software, 'Show me the molecules that are unique to the two fingerprints.' And then the software provides you with two separate images of their fingerprints." Francese shows me on the screen how the mess of overlapping fingerprints can be pulled apart to form separate, clean images.

Suddenly Francese looks at her watch. "I've got to go. I'm meeting someone who works for the Hungarian government and law enforcement."

....

Perhaps the day will come when we all wear gloves to avoid leaving biological traces of our sweat behind. At the moment, there's an opposite push: A commercial demand to develop sweat-capturing technology that can report on the inner workings of our body, primarily driven by society's current obsession with self-monitoring.

Millions of people measure the number of steps they take daily using devices like Fitbit or Apple Watch as a way to track baseline fitness. Athletes—professional and amateur alike—monitor heart rates during exercise to help them achieve training goals. Women keep tabs on their body temperature to track fertility so that they can make birth-control decisions. None of these common self-surveillance strategies require the help of a scientist or medical practitioner; at most, they just require a credit card and some curiosity.

Existing self-monitoring devices rely primarily on physical measurements: Temperature, heart rate, and steps taken are acquired by a clever mix of physics and engineering; then the data are pushed to a smartphone (and, typically, a company database in the cloud).

The next milestone in self-monitoring is chemical: Consider a device in contact with your skin that queries your sweat chemistry and sends you push alerts when you've consumed too much alcohol to drive. Or what about a car with an engine that only starts thanks to fingerprint recognition and an assessment of fingerprint chemicals to make sure you're not impaired by alcohol—or THC, cocaine, methamphetamines, opioids, or over-the-counter motion-sickness remedies that can cause drowsiness?

Athletes, pro and hobbyist, might want to keep real-time tabs on lactate levels in sweat to identify whether their exercise is aerobic or anaerobic; that is, whether they are burning carbohydrates or fats. An individual's muscle cells automatically switch between aerobic and anaerobic metabolism depending on available oxygen levels. This instantaneous information could prod an athlete to dial down muscle effort or to ramp it up, depending on whether the athlete is training for a sprint or a marathon.

Or imagine a sports team where every member on the field is equipped with a sweat-monitoring patch. Standing on the sidelines, a coach is alerted from a smartphone or tablet when one of the players begins secreting a chemical marker of extreme fatigue or physical stress that turns up in sweat—a sign that it's probably time to make a substitution. Both the military and the transport industry could also make use of such sweat-monitoring devices to keep tabs on soldiers in combat, pilots on a long-haul flight, or truckers in transit.

And then there are people with diabetes. Tracking glucose in sweat is often called the "holy grail" of chemical surveillance. The least invasive option for continual monitoring of blood sugar levels still requires people to embed needles and tubing beneath their skin, underneath a stick-on patch. Measuring sugar levels without the needles, perhaps through a smartwatch that keeps tabs on sweat, could deliver a rocket boost in quality of life for people with diabetes.

All of these sweat-monitoring applications—and more—are being researched and developed. L'Oréal has announced it is launching a skin patch that uses sweat to measure pH, to help customers choose which of their skin products to buy. A stick-on device that tracks chloride levels in sweat—a diagnostic tool for cystic fibrosis—is also close to market.

But this nascent industry, as promising as it is, has already had a spectacular failure, a cautionary tale for the whole field. In 2001, a company called Cygnus Incorporated got FDA approval to launch GlucoWatch, a device marketed to people with diabetes as a way to help keep track of blood sugar levels. GlucoWatch applied a small electric current to the skin. The current sucked out interstitial fluid—that's the biological fluid sourced by eccrine glands to release as sweat. Next, the device surveyed the fluid for sugar. When enzymes inside GlucoWatch came across glucose, they produced hydrogen peroxide. The higher the hydrogen peroxide level, the higher the sugar level. This wasn't an exact measurement of blood sugar level—it was just a proxy. Diabetic individuals weren't *supposed* to use the device to calculate their insulin needs; they were supposed to use it to follow general *trends* of decreasing or increasing glucose, to alert the individual that change was afoot.

The company did not claim that GlucoWatch was a replacement for finger pricking; it said GlucoWatch could provide supplemental information about internal sugar levels. But that didn't stop the subsequent hype. "The *Daily Mail* labeled the device 'a wristwatch to ease diabetes.' The acting principal deputy commissioner of the FDA called the technology 'one of the first steps in developing new products that may one day completely eliminate the need for daily finger-prick tests,'" wrote journalist Catherine Offord. "The excitement was tangible."

Among people with diabetes, however, the excitement was short-lived. Some users got painful rashes from the device. And it was discovered that GlucoWatch assessments of sugar levels were not reliable—one study found a false alarm rate of 51%. Within a few years, GlucoWatch production stopped, and its parent company was bought and sold, then eventually acquired by Johnson & Johnson.

The keen desire for such a device to comfortably and reliably track blood-sugar levels remains, but the design challenges are formidable. Consider the simple fact that bacteria live on skin, and that many of these microbes enjoy consuming sugar as much as we do. So sugar levels must be measured right as sweat hits skin's surface layers, before sugar levels are affected by hungry microbes. That means the sensor works best with fresh flows of sweat or interstitial fluid. And therein lies another problem: Getting a flow of sweat is easily achieved through exercise, but doing, say, jumping jacks actually negates precise measurements because exercise reduces blood-glucose levels. It's like a diabetic version of Heisenberg's uncertainty principle: By trying to measure something in the system, you inherently change the system, thereby preventing accurate measurement.

This catch-22 might be avoided by using a drug called pilocarpine, which artificially opens sweat glands when it is applied to skin with an electric current. Yet using pilocarpine to monitor glucose levels would require that it be replaced constantly, which is both expensive and cumbersome. And although periodic exposure to pilocarpine is not harmful, it is not known whether frequent, long-term exposure to the drug may be dangerous to humans; high levels of pilocarpine have caused seizures in rats.

Even if developers find a way to ensure a steady supply of sweat into a glucose-monitoring device, there's still another obstacle—arguably the most challenging of all—to overcome: There's not a one-to-one correlation between glucose levels in blood and glucose levels in interstitial fluids or sweat. Usually they all follow the same trajectory—when blood glucose levels rise or fall in blood, you see the same in sweat and interstitial fluid. But for individuals whose lives depend on accurate data, devices need to be perfect, or nearly so.

. . . .

Other biomarkers in sweat are much easier to track than glucose. If a company wants to strike gold in the sweat-monitoring world, it may be wiser to start with a device that monitors a relatively stable chemical in sweat, and one that answers lifestyle questions instead of life-and-death ones. Think alcohol.

Devices that track alcohol in sweat are already on the market, although they are much more clunky than a Fitbit or Apple Watch, and the people wearing them aren't exactly thrilled to be doing so. In 2003, SCRAM Systems launched a sweat-surveillance device called the SCRAM Continuous Alcohol Monitor, or SCRAM CAM. It was soon adopted by courts and law enforcement agencies to make sure people convicted of alcohol-related crimes don't drink, especially if abstinence is a condition of parole.

The device is a bulky ankle bracelet that functions by being, effectively, a skin Breathalyzer. Even if the wearer doesn't sweat profusely, there's always a little bit of perspiration produced by the eccrine glands in the skin beneath the sensor, which is attuned to alcohol. The sensor isn't perfect, but it's pretty good if the wearer has had more than one beer's worth of alcohol: An independent study from researchers at the University of Texas Health Science Center at San Antonio found that when test subjects drank two or three beers, the device detected the consumed alcohol coming out of their sweat pores more than 95% and 100% of the time, respectively.

Its usefulness to the justice system additionally comes from the fact that alcohol levels are measured every 30 minutes, much more regularly than the daily or weekly checks from a parole officer that used to be the norm. In the past, people under surveillance could drink in between parole officer visits and sober up for their sobriety measurement rendezvous. With a device strapped on, that's not possible, given that it can report alcohol levels back to authorities through the airwaves.

SCRAM reports that 22,000 people wear their product on any given day, and there have been more than 760,000 people monitored since the product's launch. But it's the handful of celebrities—pop star Lindsay Lohan, actor Tracy Morgan, and actress Michelle

Rodriguez—who have spread the alcohol sweat monitor's existence to the masses. Rodriguez famously called the device a "VCR
Dog Tag," presumably because the pager-sized SCRAM CAM is
large and unwieldy to individuals accustomed to miniature electronic devices.

Many people might well be cranky about having their sweat monitored by officers of the state, but others might welcome a push noti-
fication from their phone when they are getting too tipsy to drive.
Although SCRAM CAM's press spokeswoman told me there are no
plans within the company to make a commercial device for the public, she noted there is keen interest from other companies: It's not
hard to imagine a miniaturized component for measuring sweat
alcohol being added to an Apple Watch or Fitbit in the future.

Meanwhile, nary a month goes by without a bounty of new pub-
lications in prestigious journals about wearable chemical-sensing
technology: Most devices are envisioned as Band-Aid–like skin
patches embedded with miniature electronic circuits that can monitor sweat chemistry.

Researchers in these fields are juggling a multitude of engineering and chemical hurdles: How to make the patch sticky enough
that it doesn't dissolve when sweat starts flowing but not irritating to the skin. How to make it both bendable and stretchable, so
that the patch is comfortable to wear over the curves of our bodies
and can adapt to different muscle conformations involved in physical activity—all without damaging the electronics. How to develop
technology with circuits that don't require an add-on battery to
function. What about the data—what's the most efficient way to
send the information to servers or a smartphone? And if you plan to
diagnose or monitor chronic diseases with a patch, what about being
able to deliver drugs simultaneously to manage said conditions?

Yet it's a matter of time before sweat-monitoring devices become
a standard add-on to the devices we carry around. And when these
devices do hit the commercial market en masse, there will be a

huge spike in the level of intimate information being acquired and tabulated by tech companies. Certainly, companies will offer ways to anonymize personal information—Fitbit and other similar apps already offer this option. But ensuring anonymity—hacks notwithstanding—requires an active exploration of privacy settings by users.

When police solved a decades-old Californian serial killer case using DNA and genealogical data freely given to a private company, some people expressed unease about the digital health-privacy implications. And then many quickly moved on. Like climate change, even those who worry about digital privacy may feel overwhelmed by the enormity of the problem. And many may feel that Pandora's box has already been opened.

Given the widespread, blasé attitude many people have about trusting their data and its security to private companies—Fitbit, Spotify, 23andMe, and Amazon are all doing very well—our society may worry less about leaving sweaty data behind than about leaving a stinky body odor impression.

But I worry it's just a matter of time before organizations start using sweat data to triage job candidates, to set prices on health coverage, to surreptitiously check for substance use, or to determine whether a parent is fit to raise a child.

7

FAKE SWEAT

Sissel Tolaas has platinum-blond hair in a choppy Cleopatra-style cut, piercing blue eyes, and bright red lipstick. Her towering height is further emphasized by high heels. In addition to being gloriously intimidating, she is also one of the world's top odor artists, employing aroma as her medium.

Tolaas gives me a scrutinizing gaze before inviting me into her apartment in Berlin, one of those airy turn-of-the-century apartments where one spacious room with impossibly high ceilings leads to another. In her studio, thousands of scent bottles are organized on wall-to-wall shelves. This is her scent palette, the odor artist's equivalent of paint tubes.

The Norwegian artist originally trained as a perfume chemist, but instead of designing the next Chanel No. 5, she uses her skill and her scent collection to create—or re-create—all manner of complex and intriguing scents. To make the smell of World War I for Dresden's Museum of Military History, for example, Tolaas interviewed veterans and historians to get the right odorous mix of mud, cadavers, horses, gun powder, gangrene, unwashed bodies, and sickly-sweet decay. When you press on a button at the exhibition, World War I spritzes out of a nozzle, a literal blast from the past. The scent is so appalling that when the first batch ran out, the museum's staff asked her to dial down the intensity for the second installment.

Tolaas has also re-created the signature smells of Kansas City,

Shanghai, and Mexico City. To produce the scent of Detroit, one of the things she sniffed for inspiration was a burning tire from an abandoned Chevy low-rider.

Tolaas is also a purveyor of body odor, and in particular, armpit sweat odor. In 2004, as part of an artist residency at MIT, Tolaas sought out psychiatrists who worked with men diagnosed with extreme anxiety—from war veterans with PTSD to men with agoraphobia. "It was during the Bush government," she says. "I was interested in the whole notion of terrorism and paranoia. I wanted to see if I could smell fear." She met 21 patients, explained her project, and promised anonymity. "After a long time of getting their trust, they agreed to deliver me their sweat," Tolaas says. She asked the men to collect their armpit odor whenever they were anxious or panicked. "One guy was really into S&M. He would get dressed up in latex and go to a club but then he would suddenly get afraid; he suffered from extreme social phobia," Tolaas says. "I went along with this guy to the club and collected his scent myself."

To capture human body odors, Tolaas relies on technology originally devised by perfumers to trap the scent of rare flowers in the wild. Always in search of exotic new scent palettes for their blockbuster perfumes, scent chemists have gone to extraordinary lengths to capture unusual odors. The chemist and scent explorer Roman Kaiser has famously hung from a zeppelin to capture the odor of rare orchids that grow in jungle canopies and only release scent at dawn or dusk.

To trap something as ephemeral as scent, Tolaas relies on a miniature vacuum that is attached by a rubber tube to a tiny glass ampule. Smaller than a pinky finger, the ampule contains material that looks like a tiny tissue. This tissue captures the scent molecules as they pass through the glass ampule so Tolaas can analyze the scent again later. The entire contraption is pocket-sized—about the dimensions of a stack of playing cards—which makes it convenient for carrying around and collecting any interesting scents that one encounters throughout the day: Put in a new collection ampule, switch on the vacuum, and start sucking in scents.

Once the odors of the anxious men had been collected, Tolaas analyzed them and then reproduced each scent using her huge palette of aromas. She also worked with MIT scientists to encapsulate the re-created armpit aromas into a scratch-and-sniff format so gallery visitors could access the scents at their prerogative. More than a decade later, the project, called "21/21: The Fear of Smell—the Smell of Fear," is still on tour around the globe, with exhibitions at San Francisco's Museum of Modern Art, Tokyo's Museum of Contemporary Art, and many others. "The reaction is different depending on where you are," Tolaas tells me. "In the US there is so much paranoia about body odor. When it was on display, people were hesitant to smell it," she says. "They asked, 'Is it safe? Is it hygienic?' They are so paranoid. It's just chemistry.

"In Korea, there was a soldier. He was there with his grandson. He started crying at guy number 6. And I said, 'What's going on?' He told me he was touched by the smell. He said it reminded him of the war and when the soldiers were fighting close together. And all the body contact. He wanted to know if he could get some. So I sent him enough for the rest of his life."

Tolaas doesn't just hang these scents on museum walls; she also has a predilection for wearing her perspiration perfume herself. "My purpose is to be a skunk. I mean, what is smell? Primarily you use smell for communication. You can use it to say, 'Come to me.' And you can use it to say, 'Leave me alone. I need my peace.'

"I'm a professional provocateur," Tolaas tells me when I ask what motivates her work. "Companies control everything with smell and taste in the entire planet. They all deodorize, camouflage. They try to cover up reality. I want to show reality. We don't cover up bad sounds or visuals. By covering up our odors, we miss so much. I say we have to use the same knowledge to show reality."

To re-create the scent of fear—or actually any scent—Tolaas

works by trial and error. "You are working to build up a scent and one drop too much [of one component] and you have to start over again." Some of Tolaas's synthetic scents contain hundreds of molecules—she has topped out at about 500 aromatic ingredients for a given re-creation.

The different anxiety sweat odors are all distinct—some more rancid, others more musky. But one commonality in all her concoctions, in addition to the goaty, onion base inherent in all human armpits, is a molecule found in cheese: "It is present in all the men—it is a molecule made from a bacteria," Tolaas explains, but declines to name it. Like the inventor of a blockbuster perfume, Tolaas is protective of her secret ingredients.

I'm left thinking that if imitation is the greatest form of flattery, Tolaas's perspiration re-creations are a love sonnet to sweat.

. . . .

Tolaas's perspiration perfume is not the only form of fake sweat around. Tiny bottles of artificial sweat circulate the globe to satisfy the demands of an artificial perspiration market. Multiple industries—forensic, textile, jewelry—rely on a steady supply of pseudo-sweat to comply with government regulations or to ensure the quality of their products. It's hard to believe that there'd be a market for artificial perspiration. Most of us produce sufficient volumes of sweat ourselves; why would anyone spend $150 for a bottle of fake eccrine or apocrine sweat?

Clothing manufacturers, for one, buy sweat mimics because they need to make sure that textile dyes don't leach out when people perspire in their products or that colors don't change or fade in high-sweat areas like the armpit. Guitar strings must not degrade from sweaty fingers. Companies that manufacture personal handheld electronics need to make sure their phones and tablets are responsive to sweaty fingers.

Meanwhile, producers of earrings, watches, and clothing

zippers—any metal objects that touch the skin—must check that
sweat doesn't leach out significant amounts of nickel from the jewelry. That's because leached nickel can cause a rash called contact
dermatitis. And nobody wants to pay for a gold-plated product that
tarnishes or rubs off in sweaty hands. The synthetic perspiration
industry offers formulations with specific pHs or special ingredients, whatever constituent in sweat's portfolio of hundreds of chemicals is likely to pose a problem for a product at hand, from S&M
gear to spectacle frames sitting atop a sweaty nose.

Crime labs also have a steady demand for artificial perspiration.
When they are trying to visualize fingerprints, forensic scientists
typically apply a reagent called ninhydrin to turn fingerprints into
a bright magenta. But what if the fingerprint does not colorize?
Could it be that there's no fingerprint there after all? Or was there a
problem with the reagents?

That's where fake sweat comes in. For every crime-related fingerprint analyzed, forensic scientists make a control fingerprint, usually with their own digits, to check that the ninhydrin is working.
But not all forensic scientists have sweaty fingers. And they need to
do so many of these control fingerprints every day, with handwash-
ing in between, that waiting for fingers to get sweaty enough to do a
control fingerprint is not an efficient use of time. Enter an ink pad
drenched with fake sweat instead of ink.

Creating a perfect sweat mimic is impossible, because sweat is
such a personal collection of chemicals. So synthetic perspiration
products typically cater to the specific needs of a particular industry. For forensic teams, synthetic sweat needs to have the right saltiness and pH, but most important, it needs to have proteins and
amino acids that will react with the ninhydrin, so that the copycat
fingerprints will turn magenta.

One synthetic sweat company, Pickering Labs, sells more than
50 different kinds of synthetic sweat products, packaged in bottles labeled "Artificial Perspiration." When I asked Pickering Labs

about the size of its synthetic sweat sales, the company's director of operations, Rebecca Smith, demurred. But she did note that "it is safe to say we sell hundreds of gallons of artificial perspiration each year." Pickering is one of a handful of companies that sell fake sweat, and compared to synthetic mimics of other bodily fluids, such as saliva, urine, or earwax, "artificial perspiration is our largest selling product category."

....

An even more lucrative artificial perspiration enterprise is the sports drink industry, which promises to replenish the valuable components of sweat lost during a heavy workout. This multibillion-dollar industry dates back to the 1970s. Before that, athletes typically replenished their bodily fluids by drinking water. Or they got creative, as documented in the delightful 1962 film *Vive le tour*, about the Tour de France.

Jaunty music plays over slightly out-of-focus footage of riders passing through sun-drenched villages. In French with English subtitles, a narrator portrays the riders as lovable rascals, serious about their sport but not devoid of naughtiness, particularly when they arrive thirsty at tiny villages along the racing route:

> Here's one of the most important moments of the Tour: the drinking raids. [The cyclists] enter a café, shoving everyone aside. It isn't quite looting but they demand and take anything: red wine, champagne, beer. Even water, if there's nothing better. Actually, they really should be drinking water.
>
> Then they're off, usually without paying. After the Tour de France, the director of the tour receives a number of bills. Sometimes they [the riders] lose two or three minutes and have to chase the pack for the next 20 kilometers.
>
> But there's a good excuse. They sweat so much. We calculated

that on a hot day during the Tour de France, a racer can lose up to 4 kilograms (8 pounds), that's 4 liters of sweat (up to a gallon). So you have to drink and drink.

While those 1960s Tour de France competitors were relieving café owners of beverages, marathon runners of the same era were completing their multihour races without *any* midrace rehydration. "Until the 1970s, marathon runners were discouraged from drinking fluids during exercise for fear that it could cause them to slow down. For some, drinking during marathon running was considered a mark of weakness," the South African sports scientist Timothy Noakes has written.

Jackie Mekler, who ran ultramarathons during the 1950s and 1960s, told Noakes that "to run a complete marathon without any fluid replacement was regarded as the ultimate aim of most runners, and a test of their fitness."

These anecdotes seem both macho and quaint. In modern times, athletes of every level are serious about rehydration. Sweat replenishment rites begin at the very moment skin begins to glisten or even before. Instead of waiting for an internal directive to drink—thirst—many are driven to preload. But if you're not thirsty, there's no need to drink vast amounts in advance of a race. Many modern-day professional marathoners finish their competitive event lighter than they started, and some actively *aim* to be lighter on the basis of the premise that their muscles can do better with less weight. Certainly you can die from dehydration. But that's only after you begin to drop below a hydration level of approximately 15%, which happens when you are lost in the desert without water for a few days—not after a typical marathon.

In fact, for optimal performance, you don't need to be completely hydrated when you pass the finish line, says Hein Daanen, an exercise physiologist at Vrije University in Amsterdam. Though research is still ongoing, he hypothesizes that being dehydrated at the end of the marathon could be "a very positive thing because then you're lighter, and then you can run better."

But how dehydrated can you be without compromising performance? "It's a big debate in the scientific community," Daanen says. Many sports scientists point to a 3% loss of body mass as a reasonably safe level of dehydration, he says, but it all depends on the athlete's unique physiology, the sport they're doing, how they've trained, and the environmental conditions of the competition.

Meanwhile, drinking too much can lead to water intoxication, a potentially deadly condition called hyponatremia, explains Tami Hew-Butler, a sports rehydration scientist at Wayne State University. Excessive water consumption can cause internal swelling of the brain. This swelling can then fatally dislodge the brain stem: The swollen brain effectively pushes the brain stem out of place. Between 1993 and 2019, five marathon runners died from hyponatremia; none died from dehydration.

Hew-Butler witnessed hyponatremia firsthand as a volunteer at the Houston marathon's medical tent in the 1990s —it's how she got interested in sports science. Luckily, those individuals survived to tell Hew-Butler why they drank so much water. "They were fearful of dehydration," Hew-Butler says. "They said they weren't even thirsty." They were drinking excessive amounts of water because they had heard the message that rehydration was crucial, not because they were listening to their own bodies.

When we truly require rehydration, our body sends out a thirst directive. Before that thirst directive is sent, we have a sophisticated water-conservation system operating, which evolved over millions of years. "One of the amazing aspects of human physiology is that exercise stimulates a hormone called ADH (antidiuretic hormone) that causes you to retain water [and salt]," Hew-Butler explains. Hydrating against anticipated sweat loss, regardless of thirst level, is a marketing myth that only benefits the commercial beverage industry. "Listen to your body," she says, "and drink to thirst."

And when you do get a thirst directive? Drink whatever you want. Water is fine. So is juice, nonalcoholic beer, or milk. Or a sports drink if you must—certainly there are a lot of marketing dollars spent trying to convince you to do so.

....

The first commercialized sports drink, Gatorade, was developed in the 1960s by a kidney doctor and his colleagues as a recovery beverage for the University of Florida's football team, the Gators; ergo the name Gatorade. The drink's ingredients were unmistakably similar to rehydration salts, the kind you take after a stretch of diarrhea to replenish lost electrolytes: water, salt, sugar, citrus flavoring. Despite the simple cocktail that most people could make themselves from items in their own kitchen, Gatorade soon became a runaway success in American football leagues, thanks in part to enthusiastic but scientifically questionable claims of medicinal properties and athletic performance enhancement.

In the 1970s, Gatorade's parent company began to eye other markets, in particular the large number of physically active people, Noakes writes, including a "new population of joggers aspiring to become marathon runners." The company relied on celebrity endorsements and catchy phrases such as "Gatorade Thirst Quencher," "Be Tough," and "Bring It" to convince people that being part of the fitness revolution was as simple as drinking its Kool-Aid.

Gatorade soon gained a coterie of competitors, including Powerade and Lucozade (the latter based on a longtime British remedy for stomach flu rehydration that was rebranded for the sports drink market).

Sports drinks were designed to taste better than rehydration salt thanks to higher concentrations of sugar, (typically) lower concentrations of salt, and added flavoring. As one dietitian puts it: Sports drinks are just salty, flat soda. It's telling that many beverage-industry analysts group sports drink and soft drink market forecasts together.

Anyone who has run a marathon can attest to the fact that a sweet sports drink delivers welcome pep after hours of exercise. But sports drinks are so calorie-laden that most average workouts done by average people barely suffice to burn the liquid energy consumed under the premise of rehydration.

In fact, the caloric heft of most sports drinks is often counterproductive for workouts lasting less than 90 minutes. That was one of the take-home messages of a longitudinal study of sport drinks research, led by Oxford University's Carl Heneghan and published in the *British Medical Journal*. Heneghan and his team examined the studies used by one beverage company to justify its branding of sports drinks as healthy options for rehydration.

Their analysis was unflinching: "If you apply evidence-based methods, 40 years of sports drinks research does not seemingly add up to much, particularly when applying the results to the general public."

Among their criticisms, the Oxford team found that sports drink studies consistently used small sample sizes (which lowers confidence about results). Many studies involved athletes in very unusual training environments and conditions—not the typical scenarios most people face during their workouts. And the data were often acquired without blind controls, a hallmark of good science.

....

But sugar aside, what about sport drinks' other *raison d'être*; namely, replacing the salt lost in sweat? There's a very good reason our sweat glands make a valiant attempt to scavenge salt as our perspiration flows out of the gland. Salt is a precious commodity; our seawater interior relies on steady salt concentration to keep all sorts of biology on track—neurons firing, the heart contracting, to name just two. We're so dependent on salt that history is rife with wars fought and expeditions taken for access to the edible crystal. Yet just as humans have a huge diversity of sweat rates, we also have a huge diversity of sweat saltiness. The sweat glands of some individuals are better than others at recovering this salt. Before it enters the sweat gland, our interstitial fluid has a salt concentration of about 140 millimolar. By the time it reaches the skin, most people's sweat has dropped to about 40 millimolar of salt. But some folks lose double or triple the amount of salt in their sweat.

You've probably seen these salty sweaters, if you're not one yourself: After a workout, the gym T-shirts of salty sweaters dry as hard as rock and can be streaked by a white crust—malleable textile is turned into a mineral hybrid thanks to caked-on perspiration salt. One such person is Eugene Laverty, a professional motorcycle racer. Sports bike racing is more than just an adrenaline high; it's an extreme workout performed in leather. Although racing suits are created with all sorts of ingenious design elements that let air slip through where possible, leather is decidedly not a breathable textile. So sports bike racers sweat *a lot.*

When he initially entered the competitive racing circuit in Ireland, Laverty felt fine. Then he began to travel to Spain for races and eventually moved to France, where the weather is hotter. That's when he began to notice that his energy was increasingly sapped. And he began to get excessive muscle cramps. "I stopped being able to do anything in my free time except lie on the couch," he says. Was he just wiped out from the toil of competition or was there something else going on? Laverty wondered if his problems might be salt-related and tried taking salt supplements, which made him feel better. Years later, he got his sweat officially tested, and Laverty discovered that he lost much more salt in his sweat than average.

He says he started drinking an extremely salt-heavy rehydration drink—with triple the salt found in Gatorade and one-third the sugar. He says his energy returned and the cramps lessened.

Sports science researchers such as Hew-Butler and Alan McCubbin point out that there's been very little published research about whether maintaining electrolytes actually improves performance or prevents cramping. "For a lot of people these claims are almost a given. But there is very little research specifically addressing those questions," McCubbin says. What has been published suggests athletes should replace lost electrolytes gradually, but exactly how slowly is still being worked out. "Say there's a triathlete or an ultra-marathon runner and they lose say 600 milligrams an hour of salt. Should they replace 600 milligrams an hour, or should they replace half of that? Double that? Should they replace anything? What difference does it actually make? This is what I'm studying now."

In the meantime, athletes like Laverty are turning to companies such as Precision Hydration, which measures salt levels in sweat and then sells athletes personalized sports drinks.

. . . .

Precision Hydration's booth sits in the middle of a triathlon show at huge conference center in the gentrified docklands of East London. If attendees tire of the cornucopia of things to try and buy—from highly engineered exercise outfits that perform well underwater, on bikes, and on the run, to expensive pomegranate extracts—they can wander over to a stage where svelte triathletes give motivational speeches. I watch a muscled swimmer give an engaging blow-by-blow of a particularly choppy open-water race as I wait in line for my sweat salinity test.

Sports scientists have occasionally put naked athletes in large plastic bags to trap their perspiration for analysis. I'm simultaneously grateful and disappointed that nowadays many researchers rely on a tamer sweat-collection method: The salt sweat test I'll take relies on a drug called pilocarpine that will be placed on a 1-inch-diameter circle of skin on my arm to activate sweating. A gel disk containing the drug is attached to an electrical circuit, which uses a small current to push pilocarpine into my skin.

I ask the guy ahead of me how the process felt. "Well, kind of like getting a tattoo," he says, which I find a tad alarming. But when it's my turn, the pilocarpine current feels more like a buzzing tingle. After a few minutes waiting for the drug to activate my glands, the company's CEO, Andy Blow, removes the drug disk and places a new disk over the sweaty area. This second disk has tiny tubing coiled up into a miniature plastic rope rug. I watch as the sweat fills up the coil, moving around and around in the tubing. Only a few microliters of sweat are needed to measure my salt concentration.

Blow pivots over to check the status of my sweat collection. I get a bemused nod of respect for the speed of my perspiration production. "I sweat a lot," I tell him proudly.

"Actually, that *is* pretty fast," he says. "But it doesn't mean your sweat is salty."

The analysis takes barely a minute to complete before Blow tells me that my salt concentrations are painfully average: "You've come out at 50 millimolar. You're solidly in the medium category," he says. "Although it might fluctuate a few percentage points from day to day.

"Compare that to me," Blow adds. "For every liter of sweat you lose, I'm losing double the amount of salt you are." Before co-founding Precision Hydration, Blow was a competitive triathlete. "In hot conditions, racing, I would fall apart. I was taking sports drinks built for the average person. But I found out my salt losses were so much higher.

"Our business is to help athletes identify where they sit on this continuum and to give them a strategy to combat salt loss. The sports drink market likes to give everyone a Medium," Blow told me, "but there's a requirement for S, M, L, XL. We offer a bunch of different salt strengths." Mainstream sports drink companies are now also getting into the market for personalized rehydration strategies. Gatorade, for example, has been working to develop sweat-monitoring patches, so that electrolyte loses can be measured in real time. And they have begun marketing dried powder packets of electrolytes to cramp-prone athletes. But here's the thing—no sports drink from any company fully replenishes the salt losses in sweat, not even Precision Hydration's saltiest product.

For many people, it's impossible to fully replace lost electrolytes from sweat with a sports drink. Putting all that salt lost in sweat into a sports drink would make it unpalatable. Think about how salty sweat tastes—no one could swallow more than a tiny sip of liquid that salty. Even when sports drinks companies mask the salt taste of their drinks with large amounts of sugar, the salt concentration is still far below that of most people's sweat. There's just no two ways about it: The majority of salt replenishment happens by eating salty food.

The question is whether giving your body a salty boost *during* the exercise via rehydration fluids is going to have much of a performance or anti-cramp effect compared to eating a salty plate of potatoes thereafter. "If you were exercising multiple days, like when football preseason training starts, twice a day, in hot weather, non-acclimatized, or if you are in a race where it's over 18 hours and it's hot, you will lose more sodium," Hew-Butler says. "So taking in salt supplements of some kind is important. But if you're just going out for a 5K or a weight workout, if a sport drink tastes better, great, but is it doing anything? Is it giving you all that much sodium? Probably not."

As journalist Christie Aschwanden concluded in *Good to Go*, a book about the pseudoscience in many athletic recovery strategies, "there's no reason to salt your water (or your beer, as one Australian researcher did in an ill-fated attempt to make a more hydrating brew)." Even if you exercise for many hours, "you will simply correct any losses via your normal appetite and hunger mechanisms. You've already experienced this if you've ever had a hankering for a salty snack," she writes. "The research suggests that knowingly or not, people naturally self-select foods that compensate for whatever salts or minerals they lost from sweat. Even if you need to replace salt, that doesn't mean you have to drink it."

Part III

....

The War on Sweat

8

A ROSE BY ANY OTHER NAME

For most of human history, we've used perfume to modify the scent of our bodies. When I say "our," I mean our wealthy predecessors: Most historical documentation reports on the lives of the rich and powerful. And the rich certainly enjoyed a perfumed past.

Depending on the era and the location, we occasionally bathed our bodies before scenting them. But humans have consistently lobbed on perfume to overwhelm our body's odors. Wearing perfume wasn't just about fooling others about our own stink. Humans have also worn perfume as protection, as a scented distraction from the malodor of others and also as a prophylactic barrier from disease, which was long thought to be spread by bad air. Perfume kept in jeweled rings and pendants could be brought to the nose when accosted by unpleasant smells; it formed an aromatic wall.

Among the most enthusiastic perfumers in antiquity were the ancient Egyptians. At the Louvre in Paris, there is a delightful Egyptian limestone relief from around 600 BCE that depicts barebreasted women harvesting lilies in a garden overrun with flowers as tall as themselves. Other women wrap the lilies in cloth and twist the textile around two sticks above an enormous vessel, likely to squeeze out the oil or water used to extract the flower's scent. Further along the carved relief, the completed perfume is presented to Païrkep, a nobleman of the Twenty-Sixth Dynasty.

The Egyptians didn't enjoy just the aroma of lilies; they wore all manner of scents: rose, cinnamon, parsley, lemongrass, myrrh. They even concocted complex perfumed mixtures such as *kyphi*, whose 16 ingredients, including raisins, frankincense, myrrh, pine, honey, wine, and juniper berries, were ground in a mortar, soaked in wine, and/or heated to produce a thick, pungently sweet paste. It was applied and consumed to combat lung and liver disease or put on hot coals as an incense to produce a smoky fragrance. "When we think of perfumes today," writes cultural historian Constance Classen, "we inevitably imagine them as liquids. An inhabitant of the ancient world would be just as likely to enjoy perfume in the form of a thick ointment, to be smeared liberally on the body, or a fragrant smoke, infusing the air with its odour. Our own English word 'perfume,' in fact, literally means 'to smoke through,' indicating the importance this method of imparting fragrance had for our ancestors."

The Egyptian predilection for perfume was likely shared with or inspired by the Sumerians, the early Bronze Age civilization situated where Iraq and Kuwait are now. The Egyptians also shared their fragrant recipes around the Mediterranean coast: At the town of Pyrgos, in Cyprus, archaeologists have uncovered a 4,000-year-old perfume factory, where workers used olive oil produced in a nearby mill to extract the scent of fragrant plants. The prodigal use of perfume around the Mediterranean is epitomized in the writer Antiphanes's description of a wealthy fourth-century BCE Greek man:

[He] steeps his feet
And legs in rich Egyptian unguents;
His jaws and breasts he rubs with thick palm oil,
And both his arms with extract sweet of mint;
His eyebrows and his hair with marjoram,
His knees and neck with essence of ground thyme.

During the Roman Empire, perfume continued to infuse all aspects of (wealthy) life. Thanks to sophisticated Roman plumbing, fragrances were sometimes released from ceiling sprinkler systems during elaborate banquets to complement different courses, and individual dishes were also spiked with perfume. There are still fragrant dishes to be found in the Mediterranean region—pastries drenched in honey and rosewater (or honey and orange water) that hearken back to bygone eras when the plates were just as perfumed as the people.

....

In addition to the famous French palace, the city of Versailles is home to a fascinating archive of historical perfume. The Osmothèque is nestled in an unassuming part of town, a short taxi ride from the castle's resplendent grounds. Visitors can sniff perfumes worn by first-century Roman Empire elite and the protective fragrances used by French thieves who robbed plague victims without succumbing to the disease themselves. They can also inhale popular plebeian scents from bygone eras.

In the Osmothèque's display room, there are wall-to-wall vessels of perfume from past centuries: bottles reminiscent of sculptures, many crafted from fine crystal or pigmented glass sculpted into elaborate birds or flowers or figurines. According to perfume historian Eugénie Briot, the bottle itself is often more costly than the liquid it contains—even today.

The Osmothèque's founder, retired perfumer Jean Kerléo, searched historical archives for ancient perfume recipes so that he could re-create them in a laboratory. Kerléo is the mastermind behind several blockbuster perfumes, including Sublime and 1000. As head perfumer for the French company Jean Patou, he also designed perfumed gifts for visiting VIPs, invented a sporty fragrance line for Lacoste, and created the signature fragrance for

Japanese designer Yohji Yamamoto. Before his retirement, Kerléo began to re-create some of the company's first perfumes for posterity because they had no odor archive of early fragrances. People were so delighted to smell the historical scents, Kerléo tells me, that he was further inspired to re-create more historical perfumes and to begin archiving existing perfumes from a multitude of fragrance companies. "Fashion designers go to textile museums to get inspired by the past. There was no place like this for perfume," Kerléo explains.

Kerléo, wearing a black turtleneck and tailored taupe sports jacket, shows me around the Osmothèque's perfume archive and laboratory, which everyone refers to as the wine cellar. That's because most of the institution's 4,500 fragrances are stored in temperature-controlled wine fridges. The perfumes are removed from their fancy bottles and placed in brown glass jars to prevent light penetration, which can degrade fragile odor molecules. To further protect these rare perfumes, the jars are pumped full of the noble gas argon so that the liquid surfaces don't have any contact with oxygen in air, which also has a tendency to destroy fragile scent molecules.

"You might find this interesting," Kerléo says, and shows me a bottle of natural beaver musk, extracted from the animal's anal scent sac—it reminds me of leather and birch wood but sweeter. Most musk present in modern perfumes is entirely synthetic, he explained, given that the natural version is too costly. Furthermore, the invasive procedure required to extract the liquid from rear-end glands of beavers, deer, and the exotic civet cat is restricted as part of animal protection efforts.

Because the Osmothèque is an archive, the institution is also allowed to stock ingredients that are now legally prohibited in some countries for use in modern perfume. One such example is natural eugenol, a clove extract that can cause allergic reactions on some skin. "We can use forbidden ingredients to re-create perfumes because people will come and smell [on a sample stick] but won't put the scent on their skin," he adds.

One of the first scents Kerléo re-created is among Osmothèque's most venerable: a first-century perfume, the recipe for which was partially recorded by the Roman intellectual Pliny the Elder. It was called Royal Perfume, because it was a fragrance worn by Parthian royalty, who hailed from the region around modern-day Iran. "So it seemed appropriate to give the exiled [Iranian] empress Farah Diba Pahlavi a bottle of this perfume when she came to France," Kerléo tells me off-handedly as he passes me a sample.

I close my eyes and inhale. It is as if I have been swept into a Catholic church's linen closet, the air plush with the odor of incense-laden textiles—a potent reminder that the Christians borrowed their religious odors from more ancient practices. Mixed into the dense odor of incense is the smell of apple crumble, as if someone had spirited the dessert into my Catholic closet. Odors of cinnamon and cardamom mix with the scent of incense so that I feel both hallowed and hungry at the same time. "Oh my," I mumble. "It's religious and delicious. I didn't expect perfume to remind me of food."

"That's actually intentional," Kerléo speculates. "During the Roman Empire, the elites never woke up early. When they did get out of bed, they went to the *therme* to bathe. Then slaves brought them clean clothes to prepare for the best part of the day: lunch. Then they ate and relaxed." Perhaps, he says, these elites wore perfume "not just to smell nice, but to remind them of the food they were about to eat, in order to develop an appetite."

When Kerléo re-created the recipe, he had to rely on his experience as a perfumer. Pliny the Elder's recipe had listed 27 ingredients but failed to mention the relative quantities to use. "It was probably so obvious to perfumers at the time how to make this that they did not require extensive instructions." Like a cook making a favorite recipe, ancient perfumers just needed a basic list so that, you know, they wouldn't forget to add the myrrh rockrose. Or the saffron, lemongrass, lotus, marjoram, honey, wine, or the three types of cinnamon from Ceylon, China, and Arabia. To source ingredients for the Royal Perfume, Kerléo had to depend on botanists to help him track down obscure ingredients—such as Syrian bulrushes and Somalian ben-nuts, which are harvested from the pods of drought-resistant Moringa trees in the Horn of Africa. Kerléo also had to learn ancient fragrance-extraction techniques—most modern-day perfumers don't macerate nuts and then set them out in the sun to mature. It took Kerléo two years to re-create the 2,000-year-old perfume recipe, which he is careful to call "an interpretation" because it required so much improvisation, however historically informed.

We sniff through some of Kerléo's other re-creations—a scent supposedly worn by a fourteenth-century Hungarian queen, which merchants peddled as an elixir of youth because she lived to the (then) unusually old age of 75 and had married a much younger man. Her fragrance capitalized on the arrival of alcohol distillation, a tenth-century Arab discovery ("al-kohol"), to Europe. Distilled from wine, alcohol was (and is) an ideal vector for many plant-extracted scents. The Hungarian queen's perfume smells of an herb garden, as if she had kept a sprig of rosemary in her bosom at all times. The perfume was also used medicinally by people with rheumatism, Kerléo says. "Wearing something based in alcohol probably warmed joints and helped with the pain." We also inhale an ad hoc version of Eau de Cologne, made by a valet for Napoleon, who ran out of his favorite citrusy perfume when he was in exile on the remote island of St. Helena in the South Atlantic Ocean. Napoleon couldn't get a refill from *the* Cologne, Germany, thousands of miles away. Daily ablutions die hard, especially in exile.

"I would guess you are also interested in the perfume worn by the four thieves," Kerléo says. As the story goes, sometime during the plague era (which traversed the thirteenth to the seventeenth centuries), four notorious thieves stole jewelry and money from people who were sick or dying from the infectious disease. What made the particularly unscrupulous thieves famous—apart from the volume of valuables stolen—was that they managed to steal from victims without succumbing to the plague themselves. When they were finally caught by authorities, the thieves were condemned to death. But they were given a choice, Kerléo explains. "Death by torture, or a rapid death if they shared their secret for avoiding infection." That secret was the perfumed mixture they wore during their macabre activities, *le vinaigre des quatre voleurs*, or "Vinegar of the four thieves." Kerléo hands a sample over to me.

The sharp odor of vinegar makes my nostrils tingle. Along with the slight burning sensation, I am hit with a strong scent of fresh mint and other green herbs. The plague defense smells like a delicious salad dressing. I laugh. "How could this protect anyone from the plague?" It does seem unlikely, Kerléo responds, "but you can imagine that vinegar might have some antiseptic properties."

But of course! Wearing vinegar is an ingenious way to overwhelm any miserable odor—be it your own or the stink of bodies you are burglarizing. Even today, many people turn to concentrated vinegar as a household cleaner and air-freshener. It achieves both these ends for the same reason: because vinegar can kill bacteria, odor-causing or otherwise. The thieves were clever in their formulation, even though it is unlikely they could have explained why it might have worked. Several centuries would pass before scientists discovered the existence of small microbial life-forms and their role in infectious diseases. But the thieves were prescient: Disinfectant vinegar in their perfume might have killed the plague pathogens, but more importantly, disinfectants would be key to the future of modern-day anti-stink technology.

I suddenly feel like my nose has hit peak perfume. It's as if my olfactory system is an unused muscle that has just been asked to dead-lift 200 pounds. My sense of smell isn't just tired, it's lying on the ground, limp and unresponsive. Suddenly I understand why Kerléo had demurred when I asked him what perfume he wears: "I keep the scents on my person to a minimum," he had said. If you spend your days working with odors, it's unwise to overtax your own nose with odors on yourself. I thank Kerléo for the aromatic time-warp, step out into the crisp winter air, and inhale slowly and methodically. The olfactory silence is pure relief.

....

The powerful ancient scents re-created at the Osmothèque were primarily the possessions of rich elites—the common people would not have had the money to pay for such decadence. But then everything changed: Perfume became a commodity for the masses thanks to both the Industrial Revolution and scientific discoveries, which dramatically lowered the costs of perfume production. Soon anyone could improve on their personal pong with a dab of cheap perfume.

Before the Industrial Revolution, much of the perfume-making process had required careful human labor. But thereafter, pleasant odors could be extracted or built up en masse, faster and more efficiently thanks to machines and industrial steam. For example, consider the ancient process called *enfleurage*, a delightful French word that suggests being imbued by flowers. In practice, fresh flowers were exposed to grease or oil to imbue their fragrance into the liquid. Thanks to an industrial invention called the rational saturator, some 800 kilograms of hot grease could be processed in this way per day. Enfleurage could take place in one day, rather than 35 days.

Perhaps the most significant technological advance in perfume production took place in test tubes, as synthetic chemists figured out ways to produce replicas of some odors in a lab instead of extracting them from scented crops grown on farms. The development of so-called synthetic scents meant perfumers no longer had to worry about mercurial weather during growing seasons, or perfectly timed harvests, or the delicate extraction of desired odor essences by distillation or enfleurage or cold extraction (which protects the most fragile of pleasant odors easily destroyed by heat).

Instead, perfume makers could rely on a chemist to reproduce the dominant (or top) note of an odor—in a lab, often at a fraction of the cost and time.

Consider the case of vanilla, a popular perfume component. The bean and its scent exploded in popularity in the sixteenth century, after Spanish invader Hernán Cortés purportedly witnessed the Aztec emperor Montezuma drinking a chocolate beverage flavored with it. There was so much demand for the scented bean that European colonial powers tried to establish vanilla plantations around the world—initially to no avail. The vanilla orchid, as it turned out, required the presence of Mesoamerican bees for pollination to take place. In 1841, a boy on the French island colony of Réunion hand-pollinated the vanilla orchid by coercing the pollen into individual flowers with a stick. Hand pollination was a boon to the vanilla industry, but it was still an arduous process, and thus a relatively expensive perfume ingredient. In the 1870s, chemists figured out how to build a dominant component of real vanilla in a lab, a molecule called vanillin, thereby revolutionizing its role in the perfume business.

The price of vanillin dropped by 99% between 1876 and 1906, and an explosion of vanillin-based products, food and fragrance, followed. Neither the perfume producers nor their clients seemed to mind the artificiality of these new scents—even though anyone who has experienced real vanilla knows that there are a lot of subtle elements that round out the odor and flavor to provide a fabulous *je ne sais quoi*. The new synthetic vanillin made an appearance in Jicky, a Parisian fragrance launched in 1889 that soon became a worldwide hit for all genders. "Lest anyone think that unisex perfumes are a modern invention, this one was worn by both women and men ten years before an electric car, the Jamais Contente, broke the world speed record and hit 100 kilometers per hour," wrote Luca Turin in *Perfumes*.

Technological advances made perfume cheap enough for the

lower classes to buy. But as the common folk began wearing a famous fragrance, it often lost its cachet for the bourgeoisie. Case in point: synthetic heliotrope, named after a flower with a vanilla–cherry aroma. When the scent was first made by chemists, heliotrope was one of the most desired synthetic fragrances in Paris. But as the upper class abandoned the scent, its price dropped and it could be more widely adopted by the masses.

This led to a financial conundrum for the perfume industry: Should they aim for volume by selling to the masses? Or should they present themselves as luxury products and sell for high prices to the rich? Marketing turned out to be the solution. The plebs could buy scents in bulk at bazaars while similar aromas were put in beautiful bottles with attractive color-printed labels and sold at hiked-up prices to the elites at exclusive boutiques. The cachet was not entirely in the scent itself, but in its packaging and marketing.

9

ARMING THE ARMPIT

Luckily for Edna Murphey, people attending an exposition in Atlantic City during the summer of 1912 got hot and sweaty. For 2 years, the entrepreneurial high-school student from Cincinnati had been trying unsuccessfully to promote an antiperspirant that her father, a surgeon, had invented to keep his hands sweat-free in the operating room, particularly during humid summer months. As a teenager, Murphey had tried her dad's liquid antiperspirant in her armpits and discovered that it thwarted wetness and smell. She named the antiperspirant Odorono (*Odor? Oh No!*) and started a company to manufacture and sell it.

Borrowing $150 from her grandfather, she rented an office workshop but then had to move the operation to her parents' basement when her team of door-to-door saleswomen couldn't pull in enough profit. Drugstore retailers either refused to stock the product or returned bottles of Odorono, unsold.

At the time, just talking about sweat was taboo. "This was still very much a Victorian society," explains Juliann Sivulka, a historian of advertising. "Nobody talked about perspiration, or for that matter, any other bodily functions, in public."

In the early 1900s, thanks to indoor plumbing, most people's solution to body odor was to wash regularly with soap and water, and then to overwhelm any emerging stink with perfume, cologne, or diluted vinegar. People concerned about avoiding the look of sweaty armpits on clothing wore dress shields: cotton and rubber pads sewn into the armpit areas of clothing to trap perspiration.

Early sweat entrepreneurs thought they could do better. They tried peddling everything from baking soda to cayenne pepper and formaldehyde as ways to combat the odor and wetness of perspiration. One of the first US patents for a deodorant went to Henry D. Bird of Petersburg, Virginia, in 1867. It was based on disinfectants already used to clean medical equipment. If ammonium chloride, bichromate of potash, and chloride of lime could sterilize hospitals, then why not armpits? "I am aware that the following above-named chemicals have been used separately as disinfectants, but the combination named, and in the proportions given . . . is of great value in counteracting . . . the odor from the human body," Bird wrote in his patent.

Another hopeful patent-seeker, George T. Southgate of Forest Hills, New York, thought deodorants should be made of baker's yeast. There was method to his madness: Southgate postulated that the yeast would outcompete odor-causing bacteria. As he explained in his patent, the deodorant's action would be based on "the greater vitality of the yeast fermentation than that of putrefaction." In short, Southgate's idea was to sell everyone a yeast infection.

The first trademarked deodorant was launched in 1888 and was called Mum. It relied on the antiseptic zinc oxide to destroy armpit bacteria and thus prevent them from turning sweat into stink. The first trademarked antiperspirant, Everdry, followed closely on Mum's heels in 1903. It used aluminum chloride to clog up sweat pores, effectively cutting off the food source to armpit bacteria—the same strategy and active ingredient Murphey would use in Odorono. Most formulators of deodorants and antiperspirants also slipped some good old perfume into their products as a backup strategy should armpit bacterial populations bounce back or sweat pores become unplugged.

By the time Murphey started peddling Odorono, there were

already advertisements for products like Coolene, which promised to be a "medicated toilet delight," or Odorcide, which claimed it could help with "imperfections in your toilet" because "nothing is so repellant as the odor which comes from perspiration excreta."

Armpit entrepreneurs had convinced themselves that anti-sweat products could be the next big thing, and they had convinced patent and trademark authorities that their ideas were novel. But they were failing to entice the general public. Most people—if they had even heard of the anti-sweat toiletries—thought these were unnecessary, unhealthy, or both.

In 1912, Edna Murphey was learning this lesson firsthand as she tried to sell Odorono at the Atlantic City exposition. Initially, sales were so slow, it seemed Odorono would be another failure. "The [exhibition] demonstrator could not sell any Odorono at first and wired back [to Murphey to send some] cold cream [to sell so as] to cover expenses," notes a company history of Odorono. Luckily, the exposition lasted all summer. As attendees began to wilt in the heat and sweat through their clothing, interest in Odorono rose. Before long, Murphey had customers from across the country and $30,000 in sales to spend on promotion.

By 1914, Murphey had made enough money to hire professional advertisers, and she chose New York City–based agency J. Walter Thompson. She was paired with James Webb Young, a copywriter recently hired at the company's office in Cincinnati, where Murphey lived. Young had previously been a door-to-door Bible salesman. Although he had a high school diploma, Young had no advertising training—in fact he had only gotten the copywriting position thanks to a childhood friend from Kentucky who happened to be working at JWT and got him the job. But his subsequent success eventually justified the hire. Young would become one of the most famous advertising copywriters of the twentieth century, using Odorono as his launching pad.

Young first had to overcome some major obstacles: Although

Odorono stopped sweat for up to 3 days—much longer than modern-day antiperspirants—the product's active ingredient, aluminum chloride, had to be suspended in strong acid to remain effective. (This was the case for all early antiperspirants; it would take a few decades before chemists came up with a less corrosive formulation.)

The acid solution meant Odorono could irritate sensitive armpit skin. In 1914, laboratory scientists working on behalf of the American Medical Association launched an investigation of Odorono's chemical makeup, eventually calling it a "violent irritant" and "a dangerous 'perspiration preventive'" in their report, even as the company claimed on labels that Odorono was "guaranteed by the manufacturer to be absolutely harmless."

To add insult to injury, the Odorono solution was red, so it could also stain clothing—if the acid didn't eat through the textile first. According to company records, customers complained that Odorono ruined many a fancy outfit, including one woman's wedding dress, in addition to causing armpit inflammation and burning. To avoid these problems, Odorono advised its customers to avoid shaving prior to use and to swab the product into armpits before bed, allowing time for the antiperspirant to dry thoroughly and to penetrate into pores to form the plugs that limited perspiration.

Given the problematic side-effects, the company needed to convince women that Odorono was still worth their while. So Young's early Odorono advertisements presented "excessive perspiration" as an embarrassing medical ailment in need of a remedy. People were used to suffering some discomfort when taking a medical remedy; by presenting Odorono as a solution to a biological problem, Young hoped they would overlook the rashes and ruined clothing.

Odorono sales doubled, and soon the antiperspirant was being shipped as far as England and China. But by 1919, Odorono's market upswing had flattened, and Young was under pressure to do something new and better or lose the Odorono contract.

And that's when he went radical and in doing so launched his own fame. A door-to-door survey conducted by the advertising company had revealed that "every woman knew of Odorono and about one-third used the product," Sivulka says. But two-thirds of women felt they had no *need* for it.

Young realized that improving sales wasn't a simple matter of making potential customers *aware* that a remedy for perspiration existed. It was about convincing two-thirds of the target population that sweating was a serious problem in need of a remedy. So Young decided to present perspiration as a devastating social *faux pas*, one that made your presence unpleasant—even if you might not notice the stink yourself; one that people would gossip about endlessly; one that would doom you to eternal unpopularity.

His advertisement in a 1919 edition of *Ladies' Home Journal* didn't beat around the bush. "Within the Curve of a Woman's arm. A frank discussion of a subject too often avoided," announced the text below an image of a romantic situation between a man and a woman.

"A woman's arm! Poets have sung of its grace; artists have painted its beauty. It should be the daintiest, sweetest thing in the world. And yet, unfortunately, it isn't, always."

The advertisement went on to explain that women may sweat offensively and not even know it. The take-home message was crystal clear: If you want to keep a man, you'd better not be stinky.

The advertisement caused shock waves in 1919 society. Some 200 *Ladies' Home Journal* readers were so insulted by the advertisement that they canceled their subscription. According to Young's memoir, women in his social circle stopped speaking to him, while female colleagues told him "he had insulted every woman in America."

But the strategy worked. Odorono sales rose 112% within a year, to $417,000. And by 1927, Murphey saw her company's sales reach $1 million. Thereafter she sold the company to Northam Warren, the makers of Cutex, who continued using the services of JWT and Young to promote the antiperspirant.

If Odorono's early advertisements were offensive, they paled in comparison to later campaigns: "Beautiful but dumb. She has never learned the first rule of lasting charm," reads one 1939 Odorono headline, which depicts a morose yet attractive woman who does not wear the anti-sweat product.

Meanwhile it occurred to advertisers that women might stop buying their products once they had scored a spouse, and so copywriters hunkered down to address the potential loss of wedded bliss. "Why will so many married women consider themselves so safe?" begins a 1936 Odorono advertisement:

> Is it that they are blind—or just indifferent—to the secrets of appeal which single girls know so well? . . . "And so they lived happily"—all stories used to end that way. But now they begin at this point. Marriage is the stage-setting—not the ending of the play . . . Is there such a thing as being safely married?

The financial success of Young's strategy to exploit female insecurity was not lost on competitors. It didn't take long before other deodorant and antiperspirant companies began to mimic Odorono's so-called whisper copy, to scare women into buying anti-sweat products.

. . . .

Deodorants and antiperspirants initially targeted women, but it did not take long for companies to begin to be more gender inclusive given that, well, men stink too. Body odor had long been considered a fact of being masculine. It remained a cultural norm nobody particularly wanted to topple until deodorant and antiperspirant companies realized that by doing so they could double their market—and presumably their profits.

Copywriters "began adding snarky comments at the end of advertisements targeted to women saying, 'Women, it's time to stop letting your men be smelly. When you buy, buy two,'" says Cari Casteel, a historian of technology and medicine at SUNY Buffalo. That's because many men didn't want to buy such products themselves. Case in point: When J. Walter Thompson, the advertising company responsible for Odorono's success, surveyed its own employees in 1928 to see whether any male employees might consider wearing Odorono, one man replied, "I consider a body deodorant for masculine use to be sissified."

But the potential market was not lost on everybody. One JWT employee answered: "I feel there is a market for deodorants among men that is practically unscratched. The copy approach is always directed at women. Why not an intelligent campaign in a leading men's magazine?" By 1935, the American company Corcoran launched a deodorant specifically for men; they put it in a black bottle, named it Top-Flite, and sold it for 75 cents.

As with the products for women, advertisers preyed on men's insecurities. Instead of focusing on unattained romance, copywriters for male products poked at a different insecurity: Thanks to the 1929 stock market crash, 1930s men were worried about finding or losing a job. Advertisements told men that being stinky in the office was unprofessional and embarrassing, a grooming blunder that could foil your career.

"The Depression shifted the roles of men," Casteel says. "Men who had been farmers or laborers had lost their masculinity by losing their jobs. Top-Flite offered a way to become masculine instantly—or so the advertisement said." To do so, the products had to distance themselves from their origins as a female toiletry.

In the 1940s, the makers of SeaForth deodorant sold some of their products in ceramic whiskey jugs, "because the company owner Alfred McKelvy said he 'couldn't think of anything more manly than whiskey,'" Casteel says. People in sales were advised to develop "a special vocabulary" for men's products, "using adjectives like tantalizing, crisp, suave, vigorous, robust, virile, and manly." In 1965, *Life* magazine reported on the "Big Boom In Men's Beauty Aids." That year 20% of the US cosmetics market was spent on items for men.

After anti-sweat entrepreneurs poked at the insecurities of adult women and men, they turned to targeting teenagers, as part of a ploy to make these products central to America's grooming routine from puberty onward. A multitude of deodorants and antiperspirant brands flooded the marketplace, with names such as Shun, Hush, Veto, NonSpi, Dainty Dry, Slick, Perstop, and Zip.

....

Yet the success of deodorants and antiperspirants—their meteoric rise to form a $75 billion industry—is not solely due to clever marketing. The products themselves improved, evolving away from their origins as greasy or corrosive products that were gross or awkward to apply.

For antiperspirants, the biggest problem to overcome was their acidic base, which ate through clothing and caused skin rashes. The acid was essential for stabilizing aluminum chloride, the antiperspirant ingredient that migrated into pores where it crystallized to create a plug. This chemical marriage between a metal (aluminum) and a salt (chloride) was precarious; they would not remain a functional duo unless suspended in hydrochloric acid. Without the support of a strong acid, the aluminum and chloride would precipitate out into a solid powder that was of no use for combating sweat. But strong acids are not gentle chemicals.

By 1939, Jules B. Montenier, a Chicago-based chemist, had hit upon the idea of adding a third molecule that could support the tenuous association of aluminum with chloride. He discovered and patented a suite of buddy molecules, mostly involving nitrogen atoms, that, when present in the formulation, could support aluminum chloride's union, so that the solvent no longer needed to be extremely acidic. Thereafter, the solvent only needed to be *rather* acidic, reducing the likelihood that customers would damage their wardrobe and skin in the quest for sweat control. Montenier called the new antiperspirant Stopette.

Montenier also developed a new dispensing strategy for Stopette, a design that addressed a common complaint from people who used other popular antiperspirants on the market. Instead of being dabbed onto armpits with a sponge or a cotton ball (like Odorono) or being rubbed in with fingers (like Arrid Cream), Montenier patented a plastic squeeze bottle dispenser. Antiperspirants could now be "misted" onto the underarm region. By the early 1950s, Stopette was selling millions of bottles (if you believe the vintage television advertisements).

Meanwhile, another cosmetics company, Elizabeth Arden, was coming up with an even better way of formulating aluminum salts. Now the active ingredient didn't need to be suspended in a *rather* acidic solution; it only needed to be suspended in a *slightly* acidic solution, one with a pH of 4, about the same as the sweat seeping out of your pores.

This new active ingredient, aluminum chlorohydrate, was, according to the industry historian Karl Laden, one of this market's most important technological breakthroughs. And it remains one of the most common antiperspirant ingredients in the market today. If you are wearing antiperspirant now, it is most likely that aluminum chlorohydrate is the compound clogging your sweat pores. Yet the major downside of aluminum chlorohydrate compared to aluminum chloride is that sweat pores are not clogged with the same efficacy. Original Odorono and Stopette solutions could keep pits dry for days. Yet by being so much less irritating to the skin, the new antiperspirants containing aluminum chlorohydrate could be worn daily, turning antiperspirant from a product worn on special occasions to an established member of one's daily ablutions.

One major problem remained: Anti-sweat products took ages to dry, whether you spritzed them on with a spray bottle or rubbed them in with your fingers. After application, you'd have to walk around with arms raised for several minutes before putting on a shirt. This was a problem a chemist named Helen Barnett decided to fix. In 1952 she was working for Bristol-Myers, which owned the deodorant Mum at the time. Barnett was part of a team of product developers working on reformulation of toiletries and drugs. But her most prominent project was developing the first roll-on deodorant, after being inspired by the ballpoint pen, another device that minimized over-dispensing of a liquid product. The first roll-on deodorant prototype was called Roulette, and it failed miserably for technical reasons. (Gossip had it that armpit hairs were getting stuck in the prototype rolling ball mechanism.) But the beta version, called Ban, was a hit.

Roll-ons weren't the only invention to get around the problem of a drippy product. This was accomplished in another way in the 1950s thanks to the development of aerosol sprays, a technology that exploded out of the starting gate and then crashed in the 1970s.

We have the US Department of Agriculture to thank for bequeathing us aerosol spray cans, which they patented in 1941 for dispensing insecticidal bug spray. According to industry historian Laden, "the aerosol format for underarm application was an immediate success. Not only did many roll-on and cream users switch to aerosols, but the aerosol form began to attract many men into the category who previously did not use underarm products." By 1973, aerosols made up more than 80% of the deodorant and antiperspirant market.

But our love affair with aerosols peaked, as an alarming list of problems emerged: Placed too near a stove or heater, pressurized cans could explode. Teens started huffing to get high. And then there was the ozone layer: Environmental scientists identified fluorocarbon propellants (the molecules that help eject the product) as responsible for depleting the ozone layer. By the mid-1970s, federal agencies began to consider how to better regulate the problematic propellants.

Another problem for aerosols: Sometimes there were unintended health side effects. For example, the propellants and pore-clogging ingredients present in antiperspirant sprays were not just landing on armpits; they were also getting accidentally inhaled. In 1973, soon after Gillette introduced two new antiperspirant aerosols, the company withdrew the products. According to *Changing Times: The Kiplinger Magazine,* "The company discovered, and told FDA, that monkeys exposed to the sprays developed inflamed lungs."

Even though company scientists eventually developed safer ingredients that didn't deplete the ozone layer or damage lungs, aerosol-delivered deodorants and antiperspirants never recovered their popularity in America, although they still have a solid market share in Europe.

．．．．．．．．．．．．

Amid the ingredient lists on anti-sweat products, few components inspire as much public anxiety as aluminum, which is found in all antiperspirants. Although it's possible to block armpit *stink* without aluminum, the only products that can control *wetness* work by blocking sweat pores. And the only market strategy for making a sweat-pore plug involves aluminum.

Like lead, aluminum has no biological role in our bodies. That's in contrast to iron, copper, and zinc—metals that we need, at least in small amounts, to shuttle oxygen to our extremities, fight pathogens, heal our wounds, and keep our insulin levels in check.

We don't need aluminum, yet it is found everywhere on the planet: Aluminum is one of the most abundant metals in Earth's crust. And because there's substantial aluminum in bedrock, the metal percolates into water supplies around the world. Thereafter, it is taken up by plants and appears in much of the food we eat: Sesame seeds, spinach, and potatoes all harbor relatively high amounts of aluminum in their tissues, as do tea and some spices, including thyme, oregano, and chili powder. Some processed foods also include aluminum ingredients—such as sodium aluminum phosphate and sodium aluminum sulfate—as stabilizers.

Aluminum's abundance on Earth, in our water, and in our food means it is also present in our bodies, and it always has been. That's one reason our kidneys evolved to clear it and other toxic compounds out of our bodies. Much of the aluminum we ingest in food passes right through and gets pooped out. But when aluminum is absorbed by the gut, our kidneys filter it out and expunge it in pee. Even so, some of the metal stays inside. Healthy humans typically have 30 to 50 milligrams of aluminum kicking around for every kilogram of body weight, congregating primarily in lungs and bones, but also in intestines, lymph nodes, breasts, and the brain. To avoid pushing our aluminum body burden beyond what our kidneys can handle, the World Health Organization suggests that individuals limit their consumption of aluminum to 2 milligrams per kilogram of body weight per week. But the point is this: A body burden of some aluminum is the unfortunate reality of living and eating on Planet Earth.

And yet we don't need the metal. And in high amounts, it can pose a serious neurological hazard—that's why we evolved ways to expunge it. For people with weakened kidneys, for example, aluminum poisoning can be a serious concern. In the early years of kidney dialysis, some patients were accidentally poisoned with aluminum that wasn't cleared out of their system; they exhibited memory lapses, paranoia, psychosis, muscle incapacity, convulsions, and death. Because people poisoned with high levels of aluminum exhibit some neurological characteristics similar to dementia, researchers have wondered—and checked—whether the metal causes Alzheimer's disease: The best evidence suggests it doesn't. Many studies have disproven this theory after it was initially proposed in the 1960s and 1970s. But the theory endures, so much so that the US Alzheimer's Association and other patient advocacy groups find it necessary to post prominent explanations on their websites to clarify that "studies have failed to confirm any role for aluminum in causing Alzheimer's. Experts today focus on other areas of research, and few believe that everyday sources of aluminum pose any threat."

Still, too much aluminum is not good for the brain. The question is this: Given that we all ingest some aluminum in our bodies from the food that we eat, does using aluminum-laden antiperspirants push a person's body burden of the metal beyond safe levels?

The most accurate answer is no: According to the best available evidence, assessed in 2020 by European risk authorities, using aluminum antiperspirants does not pose a risk to your health. But there's a caveat: There's precious little research on exactly how much antiperspirant aluminum passes through the skin and into the body. Despite the presence of aluminum in personal care products for more than a century, there have only been a handful of research endeavors that have specifically tracked the metal's body burden through human skin—three at the point of publication of this book. For making any kind of scientific conclusion, that's a slim suite of research.

In contrast, there is an abundance of studies evaluating how much aluminum we absorb into our tissues from the food we eat, to determine how much of the aluminum present in food ends up being absorbed by the gut. For a long time, public-health-risk analysts evaluating aluminum exposure from antiperspirants assumed that aluminum absorption through the skin would be analogous to the metal's absorption through the gut—but they never actually *checked* whether this was a reasonable assumption.

In 2001, a first experiment examining aluminum penetration through skin took place. Scientists put the common antiperspirant ingredient aluminum chlorohydrate in a single armpit of two human subjects—one man and one woman—and followed them for 7 weeks, taking regular blood and urine samples. They found that only a very tiny amount of the aluminum crosses the skin and enters the body—just 0.012%, an exposure that is about 40 times less than that from aluminum in food. In the paper subsequently published in *Food and Chemical Toxicology*, the team wrote that "a one-time use of aluminium chlorohydrate [the form of aluminum in many anti-perspirants] applied to the skin is not a significant contribution to the body burden of aluminium." Although a comforting conclusion, a study on two individuals is not the epitome of rigorous science—a fact acknowledged by the scientists themselves, who referred to their own results as preliminary.

Clever readers might also question the validity of a study exam-ining a one-time use of antiperspirant. Most people use these prod-ucts every day. Putting on antiperspirant once and tracking the effects isn't exactly a realistic scenario. What if there's a problem-atic body burden from daily use?

Concerned by the general lack of data on antiperspirants, in 2007, France's federal agency responsible for the safety of health products requested a larger study about aluminum dermal absorption from French scientists. But studying live human subjects is expensive, so the French researchers used a proxy instead: abdominal skin surgi-cally removed from five people undergoing tummy tucks. The scien-tists stretched the skin over a receptacle containing saline solution and applied antiperspirant products—stick, roll-on, and aerosol—to the stretched skin. Then they measured the amount of aluminum that passed through the excised skin and into the saline solution. The study wasn't on armpit skin, and it wasn't on a live human whose circulatory system is working to process and metabolize the stuff blocking the pores. So the experiment wasn't particularly enlight-ening about real-life aluminum body burdens in humans.

For what it's worth, the researchers found that the amount of alu-minum crossing the dead skin wasn't worrisome—in most cases. But they did have one exception: They did an additional experiment on one of the skin samples to mimic the act of shaving. They wanted to explore whether shaving creates small cuts that can increase the absorption of aluminum across skin and into the bloodstream.

To do this test, the scientists repeatedly stuck surgical tape to the dead skin's surface and then ripped it off. (Apparently this is a common protocol to mimic shaving—but one wonders why they did not just run a blade over the skin.) When they applied antiperspirant to this proxy damaged, dead skin, the aluminum absorption rate was much higher, leading the scientists to conclude that "high transdermal aluminum uptake on stripped skin should compel antiperspirant manufacturers to proceed with the utmost caution."

The study, published in 2012, inspired the French medical regulatory agency to sound an alarm that got picked up by Norwegian and then German regulators, who all began to worry about the safety of antiperspirant products. In situations like these, the European Union's Scientific Committee on Consumer Safety (SCCS) gets called to weigh in on the matter.

After an initial evaluation, the scientists and risk assessors on the EU-wide committee concluded that the French study—the one that initially got European regulators in a tizzy—had too many scientific shortcomings to be considered in a rigorous safety evaluation. So the SCCS told the cosmetics industry that it needed to do a long-term evaluation of antiperspirant aluminum through skin in living humans.

Finally in 2020, the SCCS adopted a final assessment from experiments performed in 18 humans. On the basis of the new work—the most rigorous study to date, but also the *only* rigorous study to date—the SCCS concluded that "systemic exposure to aluminium via daily applications of cosmetic products does not add significantly to the systemic body burden of aluminium from other sources."

So in other words, you *probably* don't need to sweat the aluminum body burden from antiperspirants. After years of following this issue, I'm glad there's been *some* serious science done on the question of aluminum's travels through human skin. But the body of evidence is unfortunately slim. I'd like to see the studies repeated by multiple other laboratories—that's just how good science is done.

In the meantime, I wear antiperspirant. Not daily—but whenever I don't want any inherent anxiety to be on public display. Otherwise, deodorant suits me just fine. I approach antiperspirants as I do booze: With respect and moderation.

....

When I meet Chris Callewaert at the University of California, San Diego, campus on a hot August day, we've both got a shiny summer forehead sheen of not-quite-evaporated sweat. It's an unusually humid day for Southern California, the weather conspiring to make our sweaty *tête-à-tête* uncomfortably germane. So we decide to take shelter in the air-conditioned atrium of the university's biomedical research facility.

Online, Callewaert is known as "Dr. Armpit," a Twitter handle that aptly describes his research focus. In real life, Callewaert is so soft-spoken that you have to lean in to hear his Flemish lilt. He's fond of making armpit aphorisms, such as "there are more bacteria in your armpit than humans on this planet, so you should never feel alone."

When Callewaert performs what he calls an armpit transplantation, he swabs one person's armpit bacteria, and then transfers it into another person's armpit with the hope that the relocated ecosystem will thrive in the new environment. You might wonder why anyone on Earth might be tempted to do such a thing. The answer: to combat stink.

Callewaert's idea is as logical as it is wacky. We know that armpit microorganisms are responsible for turning sweat into a stinky aroma. And, let's face it, some people are naturally more stinky than others. Our aroma is related to our genetics, the food we eat, and the environment we live in, but the largest contributor to our odor is the ecosystem of bacteria living in our armpits. And some kinds of microbes contribute much more to armpit odor than others. If you have a higher percentage of *Corynebacterium* living in your armpits, there's a good chance that you'll produce a more potent and objectionable aroma than someone with a lower percentage.

This is one of the insights that emerged from Callewaert's doctoral work, and it's part of the inspiration for his postdoctoral research at the University of Ghent in Belgium and in the laboratory of UCSD's Rob Knight, one of the world's foremost researchers on the human microbiome. Like naturalists of old, microbiome researchers are tabulating the trillions of microscopic inhabitants living in all of humanity's nooks and crannies—from the wet, warm tropics of the mouth to the dry desert ecosystem of the elbow.

As a connoisseur of the armpit, Callewaert is fascinated by the fact that, when people stop using anti-sweat products, some begin to stink more than others. Aroma is certainly subjective, but one person's eau-de-naturel can be barely noticeable while another's is a potent smog.

Callewaert became fascinated by this dichotomy after an unexpected encounter with a woman. Growing up, Callewaert never found it necessary to wear deodorant. Even on the hot muggy day we meet, Callewaert tells me he hasn't put on any anti-sweat product. Over the few hours we spend together, I make a concerted effort to detect his body odor—and come up empty.

One day in his early twenties, Callewaert tells me, he had a romantic tryst that changed the course of his career. "I shared a bed with a girl," he says, "and from one day to the other I started to smell bad." Callewaert's armpits had been infected by his lover's armpit bacteria. The day of the tryst in question, he had made a rare decision to use deodorant—products typically containing antiseptics that kill armpit bacteria. "My armpit microbiome was down that day," he says, making his microbiome more susceptible to change.

"I first noticed it the day after—that I had a new body odor. I smelled sour, all the time. Even when I got out of the shower. It was always there. I went to the doctor. I tried to wash it away. But it was always there."

So as scientists are wont to do, he began to search the academic literature, hoping to find an explanation for his stinky conundrum. As Callewaert became increasingly convinced that he had been colonized by odor-causing bacteria from the person he had slept with, he began to wonder: If it was possible to colonize someone with potent, odor-causing bacteria, then what about the reverse scenario? Maybe it was possible to reduce an individual's stinky aroma by colonizing the person's armpits with the armpit bacteria of people with less potent pong. Callewaert went to a couple of professors with this idea. "And that's pretty much how I got into my PhD," he says.

Moving populations of microorganisms from one place to another to achieve a goal is nothing new. We've been doing it for thousands of years to brew beer, to make bread, and to ferment milk into cheese. Many people try to improve upon their digestion by swallowing probiotic pills laden with *Lactobacillus*, hoping that the good bacteria will settle down, multiply, and generally improve the gut's operations.

Callewaert wondered what would happen if he smeared microbes from the skin of one person's armpit onto someone else's. It seemed like a relatively tame resettlement: after all, we do something similar every time we shake another person's hand.

But therein lies the rub. The skin microbiome is remarkably *stable*. Most of the time, when your microbiome encounters a population disturbance, it bounces back. When we shake hands, we are briefly inoculated with a new ecosystem of somebody else's hand bacteria, but usually, our own hand microbiome swallows up the newcomers and returns the ecosystem to its status quo. You need either an invasive pathogen or a weakened immune system to disrupt the natural order.

The ecosystem of microorganisms that reside in a person's armpits depends on the unique chemistry of that person's sweat, skin, environment, and diet. It takes a pretty big disruption to modify the ecosystem, and more often than not, when Callewaert tried transplanting new armpit bacterial populations, the old ecosystem bounced right back. This was often the case even when he first cleaned the human receiver's armpits with antiseptic so that there was a clean slate for the donor transplant.

It was true for him, and true for many of the individuals whose armpits he has tried to improve with a microbial makeover. When he'd swab on new bacteria, the air-dropped ecosystem would soon be overwhelmed by a resurgence of the old. This is probably a good thing: It means most people's skin microbiome can defend them from pathogens. Certainly, some bad bugs do manage to take control on occasion—giving rise to infections or new malodor—but on the whole, most of the microbes on your skin are in cahoots with you: You keep making your unique brand of sweat and oily secretions on which they can live, and your ecosystem of microbes protects its territory—your body—from invaders.

The first long-term success Callewaert had with an armpit transplant was when he moved armpit microbiomes between identical twins. There were enough similarities in body, sweat, and skin chemistry between donor and recipient that the switch was successful. Subsequently he's had a few more successful armpit microbiome transplants between family members. As for Callewaert's own body odor problem? He managed to solve it, but not with a transplant from somebody else's armpit. He solved it with an accidental self-transplant: the residues of his own old armpit microbiome on an unwashed shirt he found.

It happened when Callewaert decided to paint his house. "I had this old painter's T-shirt with stains all over that I always used when painting." Every day he painted, he would wear the same old cotton T-shirt that hadn't been washed since the last time he had painted, before his tryst. Day after day he inadvertently covered himself with his old sweat and his old microbiome. As his painting project continued, he noticed he was smelling again like his old, pre-tryst self.

During his stinky era, Callewaert had been sampling himself. "So I had a good idea of the invasive bacteria that was living in my armpits." In this stinky state, he had a high fraction of *Corynebacteria*, which is often high in people with a potent pong. Cotton, as Callewaert later discovered, is a good platform for less-odor-causing *Staphylococcus* to grow on. And when he sampled his armpits after the painting episode, lo and behold, his fraction of *Corynebacterium* had dropped and his fraction of *Staphylococcus* had risen. "Before, *Corynebacteria* was like 50 to 60%; and then it was like on 5 to 10%."

These days Callewaert continues to develop armpit transplant strategies. He is also working on a project to correlate different armpit microbiomes with body odor perception. Armpits aren't home only to *Corynebacterium* and *Staphylococcus* but to a whole host of other microorganisms. Other inhabitants, even minor ones, can have major effects on stink. And what is true for other regions of the body—that greater diversity of species tends to be an indicator for health—doesn't hold true for armpits. In our moist, wet armpits, a greater diversity of microbes often means stronger stink, as rare bacteria such as *Anaerococcus* overcompensate for their low numbers by producing extremely potent odors.

. . . .

Callewaert isn't alone in transplanting microorganisms to the armpit as a way to combat stink. Biotech start-up AOBiome has been selling a product called AO+ Mist, which delivers live *Nitrosomonas eutropha* bacteria onto your skin with every spritz.

The business is predicated on the idea that humans lost their battle for a healthy skin microbiome 5,000 years ago, when the Babylonians invented soap. As the company claims, by using soap, humans decimated their skin populations of *N. eutropha*, which feeds off ammonia, a component of sweat. The company claims that spritzing on ammonia-consuming *N. eutropha* will lower skin pH and the populations of skin bacteria that are often associated with stink.

New York Times Magazine writer Julia Scott tried the bacterial tonic as her one-and-only toiletry for a month. Although her body odor spiked in week 2 given the absence of deodorant in her new bacterial ritual, the spritz ultimately improved her complexion. But as soon as she returned to showering with soap, the bacteria disappeared from her skin. "It had taken me a month to coax a new colony of bacteria onto my body. It took me three showers to extirpate it. Billions of bacteria, and they had disappeared as invisibly as they arrived," she wrote.

Scott's experience is consistent with Callewaert's: our bodies have stable microbiomes that can't be easily remodeled. It's hard not to be a little skeptical of the claim that *N. eutropha* was our body's natural odor-killer back in the pre-soap days. How can anybody know that? It's not as if microbial sequencing was available at the time. And any bacteria that is too weak to handle a bit of soap doesn't seem like much of a warrior for my skin's microbiome. If progress means I have to spritz myself with bacteria, who needs progress?

For some demographics, soap is more than enough of a weapon against stink. There's also an entire subculture of people who make their own homemade deodorants—recipes abound on the Internet. Often these recipes employ baking soda to trap odors, in much the same way as it is used in fridges to trap unpleasant food smells. Baking soda and other fragrant ingredients are made into a paste with coconut oil or shea butter and then rubbed into the armpit— an application technique similar to that of the first trademarked deodorant, Mum.

Or they turn to health food store products, some of which have ingredient lists that overlap with standard pharmacy products but claim to be "all natural" on the label. Some of these "all natural" products appear to claim that they are aluminum-free, yet are actually just excellent examples of greenwashing. Word to the wise: If labels devote serious real estate to claims of being specifically free of "aluminum chloride or aluminum chlorohydrate," check the small-print ingredient list: Several natural mineral deodorants contain "potassium alum," which is just aluminum in a different chemical form.

....

While some seek old-school, DIY anti-sweat strategies, others are working on new ways to design deodorants and antiperspirants. Cosmetic company scientists are eyeing the bacterial enzymes used by these microorganisms to turn mostly odorless sweat into dank aroma. If chemists could find ways to put a wrench in the machinery of those enzymes, they would not need to kill the bacteria with deodorants or cut off their food supply with antiperspirants. They would just need to foil the machinery used by bacteria to make stinky odors.

Another idea? Capture the stinky odors in tiny little molecular cages that are formulated into deodorants. It's like the nano version of a gas mask, except instead of saving the nose from a chemical weapon, it saves the nose from an odorous one.

In the War on Sweat, it's surprising that we haven't attempted to solve the human stink problem the way we have tackled so many other health issues: with a pill. We can take a pill to remedy headaches, infections, even cancer; why can't we rely on a pharmaceutical to remedy our body odor? The idea was encapsulated in a 2012 art project and TED Talk by Lucy McRae, who describes herself as a sci-fi artist and body architect. In her *Swallowable Parfum* video, synthesizer music and dramatic gurgling provide the soundtrack for stylized close-ups of model Shona Lee exuding droplets of viscous sweat that look more like lava lamp fluid than perspiration.

A woman's voice breathlessly admonishes us to "Go beyond acces-sory." The camera ricochets between images of the viscous sweat-fluid and Lee in a hall of mirrors. She slowly brings a shiny metal pill to her mouth, while a disembodied voice from the future sug-gests we "Express uniqueness."

Lee is now sweating profusely, but something is strange with the perspiration: It has a metallic sheen! The music reaches its cli-max as the woman stares directly into the camera and our narrator announces: "Swallowable perfume: A new cycle of evolution."

In her TED Talk, artist McRae explained her vision for the future: "a cosmetic pill that you eat and the fragrance comes out through the skin's surface when you perspire. . . . It's perfume coming from the inside out. It redefines the role of skin, and our bodies become an atomizer."

It was a piece of performance art rather than science, but it doesn't seem entirely far-fetched that someone would try to develop some-thing similar. If garlic can add to your personal aroma, perhaps swallowable perfume is the next frontier in our long war on sweat?

Or maybe not.

When I asked a coterie of sweat experts what they thought of the video, *scientifically speaking*, there was serious skepticism about the likelihood of being able to build a metallic molecule that was pleasantly odorous, as well as nontoxic, that could also bypass our diligent kidneys so as to circulate in high enough concentration in our blood to eventually come out in sweat. And if something you swallowed did impressively meet all these criteria, you might not ultimately enjoy the odor when it did exit your pores—and yet you would *not be able to stop the odor's exodus*. In other words, the con-sensus from everyone I asked was that swallowable perfume lies somewhere on a continuum between dubious and dreadful.

I don't debate any of this. But if I'm being absolutely honest, I'd try swallowable perfume in a hot second.

10

EXTREME SWEAT

For Mikkel Bjerregaard, voluminous sweating started when he was 11 years old. "I'd be sitting there at school, just looking out the window, totally comfortable temperature-wise, and then boom—suddenly I'd be dripping from my armpits. My T-shirt would be soaked. It was super embarrassing and super uncomfortable. All my friends teased me about it."

Most kids in elementary school haven't yet tried antiperspirant. But Bjerregaard had already cycled through several prescription-strength formulations. "They never worked. Within a few hours I'd still be soaked." Sometimes Bjerregaard would feel chilled—even cold—at the onset of a sudden sweat, when perspiration would pour out of his forehead and armpits. "I starting bringing extra T-shirts to school so that I could switch [them] out, usually about three or four times a day."

Over the years his mom took him to seek medical advice, but doctors were mostly dismissive. "They would say things like, 'Oh, he's just a pubescent boy. He's growing and has a lot of hormones. He doesn't drink enough water. Blah blah blah.' But I was on top of drinking enough water. And I knew I was different than other kids. I knew I sweat way worse than my friends. They called me 'Mikkel with the sweaty hands.' Nobody else was changing their shirts multiple times a day. I felt that doctors didn't understand how much

I was actually sweating. I got the vibe that they didn't take me seriously."

It took many years before Bjerregaard heard the clinical term for his day-to-day experience: hyperhidrosis, or *hyper-sweating*. Most people with hyperhidrosis sweat far above average in at least one of four places—armpits, forehead, hands, or feet. By some estimates, 15 million Americans have hyperhidrosis. Like Bjerregaard, many of them aren't taken seriously by family physicians. Dermatologists tend to be better informed about hyperhidrosis, but their medical advice can sometimes be cringeworthy.

Case in point: A 2019 article in the *Journal of the American Academy of Dermatology* begins by advising that hyperhidrosis patients should avoid "triggers, including crowded areas, emotional provocations, spicy foods, and alcohol."

Can you imagine going to a doctor for treatment advice and being told that you should effectively hide out at home (because: large groups), avoid relationships (because: emotions), and decline certain gustatory pleasures (okay, fine, drinking less alcohol is wise counsel writ large). But avoiding crowded spaces and emotions?

"Honestly that [advice] makes me angry," says Maria Thomas, a woman with hyperhidrosis who runs a blog called My Life as a Puddle. "Even if I were to avoid all of those things, I could be sitting at home, totally relaxed on my couch, watching TV, and still be totally sweaty."

Mundane aspects of daily life can be challenging when you sweat in large volumes. People with hyperhidrosis have difficulty holding pencils and pens: The writing implements slip through their fingers, as do cell phones, dishes, and power tools. Handling documents in an office setting can be fraught: Wet fingers encourage ink to bleed; they also dampen paper, which can make it rip or disintegrate. Going sock-less in sandals during the summer is a considerable challenge: Sweaty feet slide around, leading to blisters, and shoes have a tendency to slip right off midstride. Common formalities like hand-shaking or high-fives are anxiety-inducing encounters, instead of being socially binding.

"I still find myself apologizing for my hyperhidrosis," Thomas said. One apology came recently when she was asked to join a group that was holding hands as part of a ceremony. "It was an automatic reflex to apologize. But I have to stop apologizing for something I can't control."

One survey of people with hyperhidrosis found that 63% felt unhappy or depressed about their sweating, and 74% felt emotionally damaged from the condition. "Those with hyperhidrosis may often feel like they are not good enough, are unworthy of being touched, and like outcasts in their own bodies," explains Thomas.

Yet the social stigmatization of hyperhidrosis is centuries old. In trying to present Uriah Heep, a villain in *David Copperfield*, as unlikable, Charles Dickens gave the character hyperhidrosis: "I found Uriah reading a great fat book, with such demonstrative attention, that his lank forefinger followed up every line as he read, and made clammy tracks along the page (or so I fully believed) like a snail."

....

Medical researchers aren't precisely sure what causes primary hyperhidrosis. There's probably a genetic component: People with hyperhidrosis often have family members that also seem to sweat abundantly.

When scientists look through a microscope at the skin of individuals with hyperhidrosis, there is nothing unusual about sweat gland size, shape, or quantity. Given the absence of atypical characteristics, many researchers suspect hyperhidrosis is related to problematic signaling of the autonomic nervous system, which is responsible for unconscious bodily functions such as breathing, digestion, organ function—and perspiration.

The autonomic nervous system may be sending unnecessary or overactive imperatives to cool down, or the nerve fibers involved in this communication conduit are misfiring. Some researchers additionally propose that people with hyperhidrosis have aberrant control of emotions, but that theory largely ignores the fact that profuse sweating happens even when they feel calm and comfortable. "A lot of times doctors will say, 'Oh you just have an anxiety problem. You're sweating because you're nervous,'" Thomas says. "But in reality, those with hyperhidrosis are nervous because they are sweating."

The embarrassment of extreme perspiration can certainly trigger more sweating, in a reinforcing feedback loop. But the baseline perspiration rate is still high. Eccrine sweat glands—the ones that produce the salty fluid—usually release between a tenth and a fifth of a teaspoon of sweat per minute. Individuals with severe hyperhidrosis can sweat as much as 80 times that rate—close to three tablespoons per minute.

"For most people, there's a linear relationship between a sweating trigger, like temperature, and the amount of sweat produced. If you double the trigger, you get double the sweat," said Christoph Schick, the head of the German Hyperhidrosis Center in Munich. "But for people with hyperhidrosis, the relationship is exponential. A very tiny trigger produces an enormous amount of sweating."

Although Bjerregaard had carved out a strategy for dealing with his hyperhidrosis in high school—he changed T-shirts multiple times a day—things took a turn for the worse when he enrolled in a business program at university. Dressing up in a suit for interviews and class presentations was the norm. "That's when the sweating really started to bother me again. It felt like middle school all over again. Not in the fact that people were teasing me. But in the sense that I was dressed up super sharp—I would look really good. But I would have massive pit stains. And they would be halfway down the side of my body [through the suit]. I couldn't hide it, even with my arms down."

He could feel people staring. "My professor would be asking me a question during a class presentation. Every now and again his eyes would shoot down to my arms and then back up to my eyes. Some people get it. But others think, 'This guy is kind of gross.'"

Bjerregaard realized he couldn't get away with bringing a few extra T-shirts or extra suit shirts along for the day: He would have to bring along entire extra suits, which did not seem like a sustainable solution.

"I didn't really care as much about it from a social aspect, but from a professional level, it definitely bothered me. It would be hard to continue a professional life or a professional career with this issue. If you're meeting with a client, and they think, 'This guy has major pit stains—does he have his shit together?' That's unprofessional."

Bjerregaard isn't the only person with hyperhidrosis to feel that personal career aspirations are limited: Aspiring chemists may fear glassware and chemicals will slip through sweaty hands. Nurses may worry that a needle might slip and hurt a patient. Playing a guitar, lifting weights, working with power tools can feel off-limits.

The frustration of having an uncontrollable bodily function curtailing his life's trajectory began to weigh heavily on Bjerregaard. "I started to feel like this was just how my life was going to be. I really wanted to find a permanent solution." Which is how it came to be that Bjerregaard asked a surgeon to permanently sever the nerve fibers responsible for telling his armpits and hands to sweat.

The procedure is called an endoscopic thoracic sympathectomy (or ETS), and depending on who you ask, it is either a life-changing cure for hyperhidrosis or a life-threatening intervention with unpredictable and problematic side effects. The technique has its origins at the turn of the twentieth century, when severing nerve fibers was in vogue. Nineteenth-century anatomists had mapped how the brain and spinal cord connected to an intricate system of nerve fibers—called ganglia—that serve the body's extremities. With this new map in hand, surgeons attempted to treat epilepsy, goiters, angina, and glaucoma by snipping the ganglia that branched out from nerve fibers along the spine. These operations were dangerous, generally not curative, and soon mostly shelved as a treatment. But not before Anastas Kotzareff, a Macedonian doctor living in Geneva, reported in a Swiss medical journal in 1920 that he had successfully reduced a patient's facial hyperhidrosis by cutting her spinal nerve ganglia.

By the 1930s, the idea to cut nerve fibers to treat hyperhidrosis had crossed the Atlantic, popularized by an American doctor named Alfred Adson. Reading Adson's articles, his patients sound as frustrated by hyperhidrosis as modern-day individuals: "Patients find it impossible to work as bookkeepers or accountants or to work with delicate fabrics that require dry finger-tips. The skin over the finger-tips often becomes macerated and tender. The patients are likewise constantly embarrassed in meeting strangers, since their hands are always dripping wet and they feel compelled to apologize in offering them the customary salutation. They frequently shun the opposite sex to avoid embarrassing situations," he wrote in the journal *Archives of Surgery* in 1935.

Adson's surgical procedure involved cutting open a patient's torso from the back to access the nerve bundles responsible for sweating, the ones to be snipped; surgeons who cut into the chest from the front would have to get around the lungs to access the same nerves. The risks of these invasive procedures for patients often outweighed the benefits, and thoracic surgery to treat hyperhidrosis wasn't widely adopted until minimally invasive procedures were designed in the 1990s. That's when video-assisted endoscopy made it possible to see where to snip without having to entirely open up the torso.

Now surgeons make a small incision at the base of the armpit, insert a mini video camera at the end of a fiber-optic cable, and use the visualization technology to guide the operation. The surgeon deflates the patient's lungs and makes another incision near the pectoral muscle to insert a device that can snip, clamp, or cauterize nerve bundles associated with armpit or hand perspiration. Then the surgeon reinflates the lungs and sews up the patient.

"It takes about 10 minutes a side," says surgeon John Langenfeld in a video about hyperhidrosis surgery made by Robert Wood Johnson University Hospital. He is also the thoracic surgeon whom Bjerregaard ultimately chose to perform the operation. In the video, Langenfeld is enthusiastic about the operation's results, as is one of his former surgical patients, a weightlifter who had struggled with hyperhidrosis. "I'd like to do this [surgery] all day, because . . . the results are great," Langenfeld says. "But the biggest thing that people should know is there is compensatory sweating which could happen." And therein lies the rub. For many individuals who get this surgery, compensatory sweating appears in new, different areas of the body after the operation, sometimes in higher amounts than before.

If you interpret it literally, compensatory sweating is the body adjusting for reduced sweating in, say, the armpits, by moving the excessive perspiration to a different area, say, the chest, groin, or the feet. Another theory about compensatory sweating is favored by the hyperhidrosis physician Schick. He points out that by cutting the nerve, the communication flow between the brain to the sweat glands is disrupted. Although this cut reduces the signal to start sweating, the severed nerve also disrupts communication in the opposite direction. This means the brain may not receive feedback messages like, "Hi, brain, chest here. We've got body temperature under control. Please stand down on the sweating." In other words, when the body is triggered to sweat for a good reason, the message to stop is broken.

Most everybody who undergoes ETS surgery gets some compensatory sweating—one systematic review found that up to 90% of hyperhidrosis patients who received ETS surgery between 1966 and 2004 got some compensatory sweating below the breast line. The question is whether the patient will find it to be a worthwhile trade-off, on par with the original hyperhidrosis, or much worse. Someone who was previously too ashamed to shake another person's hand but now can reach out with confidence may find that being able to do so trumps compensatory sweating. However, some feel unhappier with the surgery's compensatory sweating than with their original hyperhidrosis. Compensatory sweating in the groin, for example, can make it appear as if an ETS patient has urinated on themself. One post-operation survey found that 4% of ETS surgical patients regretted getting the procedure; another found that 11% of patients had regrets. Some hyperhidrosis patient advocates argue that these ETS surgery regret rates are a vast underestimation. That's because compensatory sweating can worsen over time, and such surveys are often done just weeks or months after surgery.

"ETS surgery is Russian roulette," says Cath Ford, a British woman who got ETS surgery in 2011 to reduce blushing. The nerve bundles getting snipped don't just control sweating and blushing; they also control many signals needed by internal organs. "Now I have heart issues, gastro issues, anxiety, pain. I also have no way of controlling my body temperature. I am permanently hot. My skin burns all the time. Just walking up the stairs I overheat."

In 2018, an ETS patient named Alex Blynn launched a Facebook support group for people dealing with side effects of ETS surgery, which has grown to more than 2,000 people. Many in the group discuss how they deal with side effects of the surgery and try to dissuade people considering ETS surgery with impassioned posts about the problem of compensatory sweating. There's also discussion about new experimental reversal surgeries to reconstruct the nerves severed by ETS and how to drum up the tens of thousands of dollars needed to pay for it.

When Bjerregaard was deliberating over whether to get the surgery, the Facebook group didn't yet exist. But he was still nervous about the operation. "And my mom—she was freaked out." The surgeon told him about the risk of compensatory sweating and also of the risk of Horner's syndrome, where nerve damage can lead to drooping eyelids, constricted pupils, and reduced sweating on the affected side of the face. The thing was, Bjerregaard said, "I wanted a permanent solution." Many other strategies for treating hyperhidrosis are transient, fleeting fixes that need to be revisited on a regular basis.

One such temporary fix is Botox injections in the armpit, hands, or feet—wherever excessive sweating is causing a problem. The anti-wrinkle strategy is formulated from the neurotoxin of the pathogen *Clostridium botulinum*. The neurotoxin blocks the release of acetylcholine, a neurotransmitter involved in opening sweat pores (as well as the muscles involved in wrinkles). The major downside of Botox for hyperhidrosis is the same as it is for wrinkles: It's only temporary. And, like many other cosmetic procedures, it can get very expensive.

Bjerregaard considered taking prescription drugs that interrupt the neurotransmitter signals from acetylcholine involved in perspiration. But these medications also interrupt a lot of other bodily functions and can have problematic side effects: Blurred vision, dry eyes, gastrointestinal malaise, drowsiness, and dizziness, to name a few. That's because the drugs act systemically, over the whole body, and not just in the sweaty zones.

Instead of drugs and Botox to curb sweating, some people opt to microwave their eccrine sweat glands until they are destroyed—in theory without destroying other parts of the skin. The technique relies on sucking skin into a device that looks like a miniature hand vacuum. Cooling water flows over the sucked-up skin to prevent burning on the surface as microwaves are directed deeper, to the layers of dermis that hold the eccrine glands.

Yet another hyperhidrosis strategy involves regular, mild electrocution. The treatment, called iontophoresis, requires people to soak their hands and feet in water that is wired up to gently shock them. Those who have tried the procedure say the current feels like a tingle or mild buzzing. The downside is time: To see any results, people need to soak their hands and feet for up to 45 minutes, 3 to 5 times a week. The water is sometimes doped with aluminum salts, the Botox toxin, or anticholinergic drugs.

Nobody knows precisely why iontophoresis reduces sweating—maybe the current helps plug eccrine sweat glands or perhaps the current interferes with nerve signals to open sweat glands. Although the technique works for hands and feet, it's not particularly feasible for people with armpit hyperhidrosis, like Bjerregaard. It's hard to selectively submerge armpits in a water bath, although some have tried.

Which is why Bjerregaard decided to risk the ETS surgery and go under the knife.

"I noticed the effects immediately," Bjerregaard says. "I woke up and my hands were warm and dry. My hands had never been warm *and* dry. And I wasn't soaking wet from my armpits. I was surprised and amazed that it worked."

Bjerregaard's full recovery took close to 2 months. "Initially I could only lift a gallon of milk. I'm a big guy, six-feet-five. If I tried lifting anything heavier than the milk I was worried I might sever the nerve. Or I could collapse my lungs. So I took it slow."

Like almost everyone who undergoes ETS, Bjerregaard did get some compensatory sweating. But for him, it wasn't extreme. "Now my feet sweat more than before—but it's all more manageable. If I'm out on a hot day there's also more sweat on my chest. I still sweat more than most people but I can hide it better. Maybe I'm lucky."

Bjerregaard is now a manager at a car rental agency, a job that requires him to don a suit daily. He graduated university in resort management and marketing, having decided that a purely business degree didn't complement his love of sports. "Now when I play basketball, the ball doesn't slip through my hands when I try to catch it. Now the last part of my body to get wet is my armpits. I can wear a suit. It changed my life."

It's hard to reconcile Bjerregaard's positive experience with the experiences of those who have suffered severe and debilitating side effects from the same surgery. The International Hyperhidrosis Society is ultimately cautionary about the procedure. "Far too often we hear from hyperhidrosis patients about their difficult and often irreversible side effects caused by endoscopic thoracic sympathectomy (ETS) surgery," notes the society's website. "If you or someone you care about is considering ETS, serious caution and significant research should accompany the decision."

Regardless of the treatment—if any—that someone with hyperhidrosis decides to seek, I can't get over the feeling that as a society we've failed one another and especially people with hyperhidrosis by stigmatizing sweat. Unlike body odor, which some cultures are more accepting of than others, the stigma of hyperhidrosis exists around the world. People from Manila to Montreal describe shame associated with sweating too much. How can it be that this fundamental aspect of humanity is so stigmatized that people are moved to take great risks to their bodies to hide, block, or eliminate it?

. . . .

The vast majority of people who seek treatment for extreme sweating have what's called primary hyperhidrosis, the kind that starts in youth and can hit like a tsunami at inexplicable moments. There's another form of extreme perspiration, called secondary hyperhidrosis, which is a more mercurial condition. It typically appears later in life as a side effect of drugs (such as some antidepressants, insulin, and opiates, to name a few) or as a symptom of another condition, such as cancer, diabetes, heart failure, or even Parkinson's disease. Figuring out the root cause of secondary hyperhidrosis can be extremely tricky, requiring ingenious tricks on the part of doctors and patients, or just plain serendipity.

Consider, for example, the curious case reported by two Milwaukee doctors in the journal *Annals of Internal Medicine*. A 60-year-old man came in for a consultation, concerned about spontaneous sweating episodes he'd been experiencing for 3 years. About once a month, this business consultant would have clusters of about eight excessive sweating episodes: Sweat would suddenly start pouring off his body for a few minutes—and then entirely stop. Life would return to normal—for a month or so—before a sequence of sweating episodes would strike again.

His doctors were perplexed. The patient was otherwise quite healthy. His thyroid and blood work seemed normal. He hadn't traveled outside the country, so doctors ruled out an exotic disease. He was involved in a long-term monogamous relationship, so it probably wasn't an STD. Even so, the doctors tested for HIV and other diseases that might spike sweating—all to no avail.

Like the South African nurse with mysterious red sweat, the medical conundrum was only cracked serendipitously, after another office visit, when the patient had a sweating episode directly in front of the doctor. "The patient reported that he felt it coming on; he lowered his head into his hands and had slowed verbal responses for approximately two minutes. His pulse rate and blood pressure remained normal. The sweating was profuse and a puddle of sweat accumulated on the examination room desk where he had rested his elbows."

The fact that the patient's answers became slurred during the sweating episode nagged at the doctor. And that's when the light bulb went on: Slurred speech can happen during a seizure. So the doctor arranged to get the patient on an EEG machine and discovered that his sweating incidents were caused by semi-regular seizures. The same part of the brain responsible for the seizure was activating the sweating process. The patient went on antiepileptic medication, and the cyclical sweating episodes dialed right down.

....

In the annals of secondary hyperhidrosis—the kind that hits as a side effect of another condition or pathogen—there's arguably none so unusual and alarming as the deadly perspiration plague that befell medieval England beginning in the fifteenth century.

It had many monikers: the Sweating Fever, Sudor Anglicus, *la suette anglaise*, English Sweating Sickness, or simply Sweate for short. The disease was unusual in many ways: First of all, it killed people in their prime, mostly men, instead of toppling the very old and the very young, as other medieval pandemics did. Also, it struck the noble and wealthy as much as it did peasants, perhaps even more. The Sweate appeared unexpectedly and killed its victims extremely rapidly. As the nineteenth-century German physician and medical historian Justus Friedrich Karl Hecker wrote in his tome *Epidemics of the Middle Ages*, which was translated posthumously in 1859:

> For in the latter end of May, the Sweating Fever broke out there in the midst of the most populous part of the capital [London], spreading rapidly over the whole kingdom; and fourteen months later, brought a scene of horror upon all the nations of northern Europe, scarcely equalled during any other epidemic. It . . . was ushered in by no previous indications, and between health and death there lay but a brief term of five or six hours.

The Sweate killed many of its victims just a few hours after perspiration began. If you weren't dead within 24 hours, you'd probably survive. And unlike other plagues, "there was no security against a second attack," Hecker wrote. "Many who had recovered were seized by it, with equal violence, a second, and sometimes a third time, so that they had not even the slender consolation enjoyed by sufferers in the plague and small pox, of entire immunity after having once surmounted the danger."

The first time Sweate ravaged England, in 1485, soldiers were returning home from a decisive battle to end the Wars of the Roses, the one that put the Tudors in power and Henry VII on the throne. Many noblemen celebrated the victory on one day and then suc-

cumbed to the disease on the next: "Two lord mayors and six alder-
men died within one week, having scarcely laid aside their festive
robes; many who had been in perfect health at night, were on the
following morning numbered among the dead." By some estimates,
30% to 50% of people who contracted Sweate died; other historians
tout an even higher death toll, claiming only 1 in 100 escaped with
their lives.

Over the next decades, as Sweating Sickness epidemics hit
England a second, third, and fourth time, its reputation as a rapid
killer induced panic. Public business ground to a halt, the courts
were closed and, at the first hint of a new outbreak in 1528, King
Henry VIII "left London immediately, and endeavoured to avoid the
epidemic by continually travelling, until at last he grew tired of so
unsettled a life, and determined to wait his destiny at Tytynhangar," northeast of London, Hecker notes.

All manner of alarming strategies were concocted to whip—
literally—the sweating sick back to health. The Cambridge physician John Caius, who treated patients firsthand, had this treatment
advice: "Cause theim to lie on their right side, and bowe theim selves
forward, call theim by their names, and beate theim with a rosemary braunche." Although whipping a pandemic patient sounds
entirely counterproductive, Caius did have some agreeable treatment suggestions, such as bathing the patient with fennel, chamomile, and lavender. He also gave patients handkerchiefs infused
with solutions of vinegar and rosewater to sniff—presumably to
counteract the fetid odors. For those recuperating from the disease,
Caius proposed a meat-heavy menu, as well as "butter in a mornyng
with sage" and "figges before dinner."

Meanwhile, the agent responsible for causing Sweate remained
unknown. Caius thought the Sweating Sickness was caused by bad
air—consistent with medical thought of the time. Some observ-
ers wondered if the poor English diet exacerbated the illness. As
Hecker pointed out in the nineteenth century, "Flesh meats highly

seasoned with spices were indulged in to excess; noisy nocturnal carousings were become customary; and it was also the practice to drink strong wine immediately after rising in the morning." Hecker also bemoaned the country's lack of an edible greens market: There was "a total want of refinement in diet," he wrote, so much so that "Queen Catherine had pot-herbs brought from Holland for the preparation of salads, as they were not procurable in England."

Others blamed the pandemic on mystical geophysical events, such the nearby transit of comets or the activity of Mount Vesuvius. Papists loyal to the Catholic church, which had been split apart by Lutheranism, circulated a theory that the recurring Sweate was simply the wrath of God.

Modern scholars favor a microbial culprit. Researchers have floated all sorts of theories about possible causative agents, including influenza, rheumatic fever, typhus, plague, yellow fever, botulism, and ergotism—which is an infection caused by cereal fungus that leads to convulsions and gangrene in the patient.

In 2014, Belgian infectious disease specialists argued that Sweate was most consistent with a hantavirus attack, which is typically spread via rodents and has similar symptoms—though the effects are typically more pulmonary than perspiratory. They also suggested that the English Sweating Sickness bore a resemblance to yet another set of perspiration plagues that hit the French province of Picardy in the 1700s and 1800s and then spread to Germany, Belgium, Austria, Switzerland, and Italy—a deadly disease called the Picardy Sweat, which may have been responsible for taking Mozart's life.

The Belgians couldn't resist the temptation to evoke Britain's most beloved detective in their Sweate analysis: "The mystery around the origin of the English sweating sickness is perhaps best worded by a quote of Sherlock Holmes: 'Once you eliminate the impossible, whatever remains, no matter how improbable, must be the truth,' " they noted.

They also cautioned that Sweate may return. Their call to be concerned about epidemics of old parallels the cries of a London physician named Henry Tidy. He wrote a letter in 1945 to the *British Medical Journal* about Sweate: "Such is the disease which at one time was feared even above the plague. We may bear in mind Foster's warning: 'We should be unwise to regard it as necessarily a disease nearing extinction.' "

These scholars have a point. One of the concerns of climate change is that old plagues will reemerge as frozen bodies carrying ancient pathogens melt out of permafrost and glaciers, giving the microbes archived by ice a new lease to infect life on Earth. Might we face another sweating sickness outbreak as our world continues to warm?

....

If excessive sweating can be disruptive—or life threatening—so is the opposite: A complete absence of perspiration. Being sweat-free makes temperature control in warm weather a life-threatening challenge. Yet at least one individual with this condition—the Serbian performer Slavisa Pajkic—has turned his inability to sweat into a cabaret career opportunity: On stage he has been called Electro, Biba Struja, Battery Man, Electric Man, and Biba Electricity.

Pajkic was born without sweat glands. This means his skin is very dry and extremely resistant to electricity. His sweat-free skin acts like a huge piece of rubber, a characteristic that supplies him with an apparent superpower: He seemingly can't be electrocuted.

Pajkic discovered this skill as a teenager, when he touched an electric fence and did not get electrocuted.

In 1981, Pajkic withstood a shock of thousands of volts of electricity. More recently, he has upped his game to 1 million volts. In 2001, he boiled a cup of water in 1 minute and 37 seconds by passing electricity across his body—a feat that landed him in the *Guinness Book of World Records.* "Every man is born with a purpose," Pajkic told the filmmakers in *Battery Man,* a 2012 documentary about him. "It was my luck to find out that electricity cannot hurt me."

Pajkic turned his skin idiosyncrasy, and his knowledge of electronics, into a one-man performance piece: The show climaxes when he cooks a sausage via an electric current transmitted across his body. To begin the act, he stabs a sausage on both ends with two forks, one in each of his bare hands. Then he connects himself to a circuit, so that an electric current passing across his body begins to cook the meat yet does not harm him. The trick ends when the sausage splits open as it would in a frying pan. A major downside to channeling so much electricity is that he regularly loses his fingernails.

Although Pajkic can withstand enormously high voltages that would hurt or kill many others, some of his show is smoke and mirrors. As electrical engineer Mehdi Sadaghdar revealed on his viral YouTube channel ElectroBOOM, the sausage-cooking trick can also be done with voltage levels so low that even an individual with functioning sweat glands won't be harmed (as long as they have expertise with electronic circuits—don't try this at home!). The key point is that Pajkic's highly resistant skin makes it generally safer for him to do tricks with electricity because he is able to handle much higher voltages than the rest of us sweaty mortals.

Supposed superpower aside, his summers are pure misery because of the lack of sweat glands. The only way Pajkic can survive hot weather is to wear wet T-shirts and to spritz himself with water. His mother, describing Pajkic and his brother as children, said, "Both my kids are the same. Neither of them sweats, or has sweat glands. I saw it early. On a hot day [they] would scream like hell. I had to keep a bucket full of water and line them up like ducks [to be cooled down with the water]."

Now in his sixties, Pajkic is performing less and filling in the career gap by offering his services as a healer for people with maladies that range from migraines to knee pain. Pajkic discharges electricity to his patients' wounded areas via his hands using a circuit he rigged up. Some people believe he has mystical powers despite the lack of scientific evidence. Others criticize him for using his rare skin condition to peddle pseudo-remedies.

....

For most individuals, body-wide eccrine sweat glands develop in utero between week 20 and week 30. That's when a special protein triggers formation of the millions of sweat glands found in the skin. Yet just a single tweak in the DNA for this protein can result in skin that cannot sweat. This unusual genetic condition is called X-linked hypohidrotic ectodermal dysplasia, or XLHED. It's just one of a family of rare genetic disorders, called ectodermal dysplasias, that can result in an individual having few or no sweat glands.

About 1 in 25,000 people are born with XLHED, and if it is not discovered early enough, a young child can die of heatstroke.

The condition is recessive and it is found on the X chromosome, which means it typically affects males, because they only have one copy of the X chromosome. Female children, which have two X chromosomes, would need two copies of the extremely rare mutation from both parents to develop XLHED.

In 2013, a start-up called Edimer Pharmaceuticals launched a clinical trial that aimed to treat newborns with the condition. The company's idea was to inject the babies with functional versions of the protein that induces sweat-gland formation to try and make up for its absence in utero. But the clinical trial was a failure. By the time a child is born, the drug is too late to trigger sweat-gland formation. Then in 2018, researchers in Erlangen, Germany, reported in the *New England Journal of Medicine* a rather daring and successful experiment. A woman in her thirties, who was a nurse and a carrier of the XLHED gene, had given birth to a son with the condition. As her son struggled with regulating temperature, she and her partner wrestled with the decision of whether to have a second child. That's when she found out she was unexpectedly pregnant with twin boys. Because she was a carrier, there was a 50:50 chance the twins would be born without sweat glands.

At 21 weeks of pregnancy, after some diagnostic tests, doctors told her that the twin boys would most probably also have the no-sweat-gland condition. The news propelled her into action: She knew she had just a few weeks before her two fetuses ought to be developing sweat glands. Because her firstborn has XHLED, she was familiar with a German clinic that specializes in rare genetic skin diseases—it was also the site of the failed clinical trial where newborns had been injected unsuccessfully with functional copies of the sweat gland–making protein.

She asked Holm Schneider, a doctor who had been involved in the trial, whether the treatment might work if they injected the protein directly into her uterus during the period when her twins ought to be developing sweat glands. The idea was risky: As Schneider told *MIT Technology Review*, "We were hesitant. In that situation you think twice. You think more about the risks involved—three lives— but also the chances that it may bring."

It took the doctors a month to arrange a compassionate use of the putative drug and to find leftover doses of the drug from the company. And then they injected the drug—the functional version of the missing protein that triggers sweat-gland formation—into her amniotic sac. Even though it had taken weeks to orchestrate, the drug arrived within the window of time between weeks 20 and 30 where fetal sweat glands develop. Incredibly, it worked: The twins were born with sweat glands. The team tried the treatment on a second pregnant woman, also with success. Schneider is now hoping to organize a larger clinical trial.

But even if the drug proves successful in many other patients, getting the drug to market is a long shot. The return on investment for any company developing a drug is low for rare diseases because the demand is low. Furthermore, the pharmaceutical industry has also shied away from developing drugs for use in pregnancy. "Treating babies in the womb is something few companies would attempt, because it can pose risks to the pregnant woman," writes Antonio Regalado in *MIT Technology Review*. As Schneider told him, "If you wanted to make this for the patient community, and administer it once in life, the chance that it will ever be profitable is very low. Yet here you have an incurable disorder, no drug available, and here is something working—three out of three."

Given the lack of safe and reliable treatments for people existing on both extremes of the sweating spectrum, one hopes this small success, if shown to work safely on a larger scale, might find its way to the global market.

11

SWEAT STAINS

At the height of human experience—getting married, going into battle, rocketing to the moon—we tend to don special costumes, whether royal wedding gowns, military fatigues, moon-mission spacesuits. And those moments tend to trigger adrenaline, a hormone that opens the perspiration floodgates. So the garments worn during historic events, the ones we're most likely to collect, tend to arrive at museums with a sweaty signature. Case in point: NASA spacesuits. There's the sheer excitement (or terror) of being propelled at 25,000 miles per hour off one's home planet or of the audacious spacewalks taking place in direct sun at a toasty 250°F. Sometimes astronauts perform acts of extreme physical effort—think of Eugene Cernan's early spacewalk on one of the Gemini missions, when he got so sweaty that the interior of his astronaut's helmet fogged up, making it impossible to see and nearly impossible to crawl his way back to safety. Imagine the floods of perspiration seeping into the rest of his spacesuit.

Until NASA engineers perfected spacesuit cooling systems, astronauts sweated profusely into the outfit's many layers. Some of these first spacesuits got so drenched with perspiration that some metal components, such as the suit-wrist connectors, corroded irreversibly, probably from the salt in sweat. Even the suits worn for more recent trips, such as those to the International Space Station, have a perspiration problem. Astronaut Doug Wheelock, who has spent more than 178 days in space since 2001, has a mercurial relationship with his spacesuit. "It may look cool, but it's 35 years old, smells like a locker room and there's some discoloration on the inside," he told *New Scientist*.

And it's not just the outfits we've worn to space. Dig into the textile-conservation literature and perspiration problems are not hard to find. One has a sense of how frustrating old sweat can be when the normally staid writing found in this literature turns hyperbolic about underarm areas. One article is titled with the question, "The Pits of Despair?"

Textile conservators bemoan the damage done to beautiful garments from the salty sweat of those who wore them. "Silk is the most vulnerable," says Lucie Whitmore, the Museum of London's fashion curator. "In silk garments that have been worn against the skin, the dried perspiration cracks—even shatters—the armpit area." She has also seen sweat stains on corsets and other undergarments; party frocks and waistcoats from the eighteenth century onward; as well as theatrical and performance costumes from all periods. ("Performance costume is probably the sweatiest of anything we see," she says.) Spanish textile-conservation scientists have gone as far as to image and analyze tiny salt crystals in the tide lines of sweat in the armpit of a seventeenth-century bodice "belonging to one of the most privileged social classes of the European courts." Whoever wore that bodice probably did not expect that her sweat would be subject to so much scrutiny three centuries in the future.

When sweat emerges, it's often acidic, with a pH as low as 4.5. As sweat decomposes, its pH rises past neutral 7 and into the alkaline, where it then dries. If the acidic sweat hasn't done damage to the textile, the dried alkaline product probably will, by lowering the tensile strength of the textile, a problem seen especially in natural fiber fabrics. "The longer dried perspiration stays on the fabric, the higher the degree of damage," notes one guide.

Sweat can also entice tiny fabric-munching pests to feast on an important outfit. "There are also many examples where insects have preferentially eaten underarm and crotch areas," explains Jessie Firth, a conservator at the Australian War Memorial museum. "Our suspicion is that these areas are tastier to the insects due to the presence of sweat and body oils."

Although old sweat can hurt historic textiles, one benefit to clothing from the nineteenth century or before is that the pieces predate widespread use of deodorants and antiperspirants. The acidic formulations of these products in the early twentieth century ate away many textiles. Meanwhile, the aluminum salts found in modern antiperspirant formulations can combine with soap or detergent to form a discolored, brittle crust on fabric—especially cotton—that doesn't dissolve away in water.

So what can be done about all this vintage perspiration? Is there a way to preserve garments from the old sweat of people who wore them?

....

The long, brown, hangar-like building housing the Canadian Conservation Institute is in a strip mall on the outskirts of Ottawa, sandwiched between an auto parts supplier and an Irish pub. Despite its humble exterior, the CCI is where Canada's most prized pieces of art and heritage go on a curative sabbatical; it's the place where the country's 2,500 museums and 1,000 archives send their most valuable objects, so that conservators and conservation scientists can study, authenticate, and restore them.

The cavernous space is large enough to accommodate some cedar totem poles carved by Indigenous Peoples of the Pacific Northwest. The first time I visited the institute, conservators were working on a beautiful gold-and-brown silk flag from the War of 1812, the only time Canada and the United States have fought against each other. In another room I saw scientists studying centuries-old Inuit seal-

skin shoes, trying to understand the underlying chemistry of a tan-
ning process that relied on smoke and fat from animal brains.

I have returned to the institute, having heard that a nineteenth-
century party dress boasting some problematic armpit sweat stains
is on a restoration lab bench. Laid out on a pristine white table,
the embroidered silk dress is stunning. Produced around 1890,
its label reveals it was made by "Robes et Manteau," located at 287
Fifth Avenue in New York City, an address that predated the 1920-
built Textile Building now standing on the same spot. Robes et
Manteau—French for "Dresses and Coat"*—had a manager named
Isaac Bloom, who drew branding inspiration from Paris, where the
city's impeccable designs were as popular then as they are now.

The cream-colored dress features an alternating pattern of satin
and silk stripes upon which are embroidered tiny pink flowers with
foliage in several shades of green. A forest-green velvet textile decorated with a brocade of silver, gold, and copper threads borders the
collar and bodice. The petticoat is the show stopper, though, with
thick layers of tulle embroidered with white and periwinkle flowers.
Although the dress is lovely, there are tide lines in the armpit area,
evidence that a real living human wore the dress—and sweated in it.

"It was definitely a high-society gown," explains Jonathan Walford, director of the Fashion History Museum in Cambridge,
Ontario, who sent the dress to CCI. In 1890s New York, the gown
would have been worn only once—or perhaps twice among an
entirely different crowd of people, Walford says. "If you were in New
York high society, and you wore the same dress for more than two
events, people would gossip, 'Oh *that* again.' "

---

* The American label was likely misspelled in the attempt to pass as French. Presumably the label intended to be French for "Dresses and Coats" (plural) instead
of "Dresses and Coat" (singular). But the owners of the label neglected to make the
French word for coat, *manteau*, plural by adding an "x" at the end.

The short-sleeved arms suggest whatever event the wearer had attended, it had probably been in spring or summer, which means there had probably been dancing. "Sweat can be the spoiler that ruins a dress to the point where it is not really showable anymore," Walford says.

The garment had arrived in his museum's collection courtesy of a fashion collector in Toronto named Alan Suddon. After he died in 2000, his collection of vintage high-society gowns from around the world ended up in a poorly maintained storage unit for 15 years before it was auctioned off or donated to museums across Canada.

"And that particular dress—well, nobody had wanted because of the amount of work that was required to do the conservation," Walford says. "But I looked at it and thought, 'That is an amazing dress.'"

One of the biggest concerns about sweat stains is the damage to the fabric as the fluid dries. Once the fabric is degraded, there's little textile left to actually conserve. But another major issue is the color change that can occur when sweat reacts with fabrics or their dyes.

On a light-colored dress, such as white silk, yellow stains can appear over time in the armpit area, thanks to the chemical interactions of light and oxygen with the sweat's lactic acid and amino acids that have dried within the textile.

Sweat can also react chemically with the textile pigments, dissolving them so that colors bleed into one another. If the dyes are pH sensitive, sweat's transformation from acidic to basic on the skin can also lighten, darken, or completely change textile hues. "Sometimes sweat will interact with certain colors in a printed textile more than with others. You'll find, for example, that the black is all rotted out, but the lighter colors are fine," Walford tells me.

Or consider the case of a World War II wedding dress that crossed Jessie Firth's conservation table at the Australian War Memorial. Worn by five different women in the 1940s, the pretty beige dress had developed bright-green armpits. Firth finally figured out that the culprit was a copper thread used to decorate the dress. It had been corroded by the armpit sweat, producing the green you normally see on copper-plated architecture.

As I look at the New York high-society gown sitting on the CCI lab table, Janet Wagner, a textile conservator at the institute, shows me how the dress's precious metal threads of gold and silver have been oxidized by the sweat's salt, turning the cream-colored textile in the armpit area into a shoddy brown. After hearing that the color could have been a bright green, the shoddy brown seems, in comparison, a lot less intrusive. It is almost good news.

....

Textile experts have found many ways to remove perspiration stains: flushing with steam or dabbing with solvents picked for the textile in question—such as dilute concentrations of acetic acid (vinegar), petroleum spirits, or acetone (nail polish remover). But any intervention is risky: it's easy to make things worse rather than better, to create a new problem while trying to solve an old one. This is why conservators are increasingly circumspect; in recent decades, conservators have opted for a hands-off approach to objects of cultural heritage. Whereas a conservator of yore might have thoroughly washed a garment and just re-dyed any parts that got lightened, contemporary conservators often frown on the idea, arguing that the act of re-dyeing detracts from the object's history and provenance.

The idea is that conservators should not do anything to a valuable artifact that can't be reversed. This is why they often turn to synthetic sweat, adding it to a textile similar to the one in question, artificially aging the piece in, say, a warm oven, and then testing cleaning solvents on it. In many cases, especially if the garment was only stained in the armpit, a clever display technique—a carefully placed shawl, indirect lighting—is employed to obscure or to distract attention away from a sweat stain.

This conservative approach is sometimes at odds with the expectation of many museum patrons; namely, that anything on display should look perfect. As with our own bodies, there is public pressure to hide faults and evidence that we are, in fact, aging. But to me, sweat stains make an artifact *more* interesting, because such flaws speak to its life path, the moments that made its existence (objectively) interesting. Sweat *is* history.

What if Queen Elizabeth II's coronation dress had sweat stains? I would hope that no conservator would remove such important historical information about her emotional state. Or consider the sweat stains on a president's inauguration-day outfit, on an athlete's jersey when she breaks a record, on a musician's suit after his final concert, on a ballerina's tutu after an important premiere, or on a soldier's fatigues after a long-fought battle: The sweat helps tell the story of the object and is part of why the clothing is interesting.

"Certainly, there's value in showing that real people with real bodies wore the clothing," Whitmore says. "Especially if the sweat is part of the garment's story, part of the history of the object." But, as a curator, she feels bound to additional ethical considerations. "Did the person who had the garment *want* to show their perspiration? Did they want to show their raw human sweatiness? Sweating is intimate. Some people find it embarrassing. I have to consider the opinions of the object's owner and their family."

So what to do about the dress sitting on the CCI conservation table? It is highly unlikely the woman who wore the party dress gave anyone permission to have her sweat on display in perpetuity. Then again, we don't know who she is, so it's not like her privacy or that of her family or descendants is compromised.

Ultimately, the CCI conservators and Walford agreed that the dress's sweat stains should stay for another reason: Trying to remove the perspiration could cause the dress more damage. Instead, the conservators would do a bit of surface cleaning to remove the grime from having spent several decades in storage, and they'd stabilize some weak seams to prevent any further breakdown. "Have a look at this."

Wagner, the CCI conservator, points over to the dress's skirt, showing me an additional dark-brown stain on the silk that I hadn't noticed. She suspects a splash of white wine or champagne splashed onto the dress.

"Champagne and white wine, they fool you," Walford says. "You think they don't stain but with age they do. The sugar is oxidizing and you end up with these dark-brown stains that show up years later, but at the time of course you think, 'I spilled some wine. Nope, I don't see it—it must be fine.'" Perhaps our dress owner was similarly self-deluded, if she happened to notice or bemoan the wayward splash of wine at all, amid the sweaty night of dancing and fun.

"Whoever she was, wherever she was, and whatever she drank," Walford says, "she seems to have had a great time. And I'm happy to keep that evidence."

. . . .

I'd like to raise a toast to sweat.

It keeps us alive, day in and day out, and it does so in a way that is vastly more efficient, and much less objectionable, than the cooling methods used by many other creatures on Earth. I'd argue that it's much more pleasant to sweat and to see other humans doing the same than to experience an equivalent release of urine, vomit, or poop in the name of cooling.

Let's appreciate and acknowledge the fact that over the course of human history, sweat has helped us adapt to many new environments on our planet. Like the pigeon and desert dove, we're capable of surviving almost anywhere. Let's just not warm our climate to the point where this no longer holds true.

I love that sweat helps us navigate our social world, if research suggesting its role in identifying anxiety, infection, and romantic compatibility pans out. Of course, like many human relationships, *it's complicated*.

I appreciate that our sweat keeps us honest. It can reveal many of our secrets, for better or worse, through odors, damp patches on clothing, or a careless fingerprint left at a crime scene. In this era of curated personas, it is heartening that some aspects of our humanity are still irrevocably transparent (chromhidrosis notwithstanding). We just need to keep our government, law enforcement, military, employers, and insurers honest too, and make sure they don't exploit our biological transparency.

One of the things I love most about life writ large is its amusing absurdities, and sweat is such a great source of these: The fact that even though we produce incredible volumes of the fluid, there's still a market for knock-off, artificial perspiration. The fact that we pay lots of money to spend time sweating in vast quantities in a sauna or a gym only to slap on anti-sweat products thereafter—or sometimes even prophylactically.

In the heat of summer, I *am* glad there are ways to dial down the smelly consequences of my perspiration. But this inherent gratefulness—and my motivation to control my sweat's odor, its visible existence on my skin, and its unwelcome presence on my clothing—is rooted in a century of manipulation by marketers who prey on our deepest fears of social exclusion. Let's push back on that. Everybody, let's cease and desist with the sweat stigma. Perspiration is primarily just a body trying its best to do its thing, to stay alive. Wear deodorants or antiperspirants if you want to, but otherwise let's all just live and let sweat.

....

ACKNOWLEDGMENTS

....

My utmost thanks and gratitude goes . . .

To every person who told me about their perspiration, for trusting me with the sweaty details of your life.

To my incredible editor, Matt Weiland, for being a patient and wonderful cheerleader, a shrewd wordsmith, and a bountiful producer of sweat puns.

To my agents, Christy Fletcher and Sarah Fuentes, for guiding and supporting me and this project along its winding path.

To my fact-checkers, Alla Katsnelson and Viviane Fairbank, for their superhuman ability to disappear into a haystack of facts and emerge promptly with a golden needle. Any errors are mine.

To all the scientists, scholars, entrepreneurs, patients, athletes, artists, and historians, who spent their valuable time speaking to me about their work and their lives. I am deeply grateful for our enlightening conversations.

To all the brilliant people I crossed paths with at the Max Planck Institute for the History of Science in Berlin and the Science History Institute in Philadelphia, for nourishing me with accounts of their research. I benefited immeasurably from conversations with historians Etienne Benson, Donna Bilak, Lucia Dacome, and Joanna Radin. I am grateful to both institutions for the funding that allowed me to dig deeply into the history of sweat.

To John Borland, Susanna Forrest, and Aimee Male, for helping me get this book out of my head, into a proposal, and then a manuscript.

To Anett Hauser, Ian Baldwin, Christian Hackenburger, and Jesse Heilman for their help with my gonzo sweat experiments, both Gedanken and tangible.

To Kristen Allen, Sophie Boehm, Raychelle Burks, Deborah Cole, Kelly Fritsch, Beth Halford, Susan Harada, Naomi Kresge, Matthew Pearson, Nick Shaxson, and Emma Thomasson, for their wise counsel and support along the way.

To Amanda Yarnell and Lauren Wolf, for being incredible bosses who bent over backward to support this book.

To all the members of Lauryn Mannigel's Sensory Culture Club reading group past and present, for scintillating and nerdy discussions of all things smelly and sweaty. You are my people.

To all those who sent me articles about extremely niche sweat topics, especially Davide Castelvecchi, Andrew Curry, Hilmar Schmundt, Ethan Sternberg, and Volker Oschmann—this was deliciously weird and wonderful stuff I might not have encountered myself.

To those who made me sweat—from the Aufguss masters to the spin instructors and everyone in between: I am grateful for your expertise and your benevolent sadism , especially you, Micha Østergaard.

To those who sweated with me in vast quantities—especially Gaya Marina Garbaruk, Jenny Lassmann, Mathilde Saingeon, Tionette Stoddard, and Anne Tessier: Thank you for your companionship on this salty journey.

To my mom, Nikki Everts-Hammond, for nurturing my love of prose, and to her mom, the late Peggy Everts, for showing us all how to craft stories about the absurdities of life.

And finally to Jörg Emes, who has been a backbone of this book since its inception. Thank you for holding down the fort with grace when I disappeared into my writing cave and on a multitude of research trips.

This book is dedicated to my son, Quinn, whose metric for fun at gymnastics class at age 4 was whether or not he had worked up a sweat. You make your mama proud.

NOTES

INTRODUCTION

1   **"The Case of the Red Lingerie—Chromhidrosis Revisited":** J. Cilliers and C. de Beer, "The Case of the Red Lingerie—Chromhidrosis Revisited," *Dermatology (Basel, Switzerland)* 199, no. 2 (1999): 149–52, http://www.ncbi .nlm.nih.gov/pubmed/10559582.

2   **sweat that has turned green, blue, yellow, brown, or red:** Cilliers and de Beer, "Case of the Red Lingerie."

5   **$75 billion a year:** M. Shahbandeh, "Size of the Global Antiperspirant and Deodorant Market 2012–2025," Statista, accessed July 14, 2020, https:// www.statista.com/statistics/254668/size-of-the-global-antiperspirant-and -deodorant-market/.

6   **"Far more than one . . . sexual attraction and masculinity":** Michael Stolberg, "Sweat. Learned Concepts and Popular Perceptions, 1500–1800," in *Blood, Sweat and Tears: The Changing Concepts of Physiology from Antiquity into Early Modern Europe*, ed. Manfred Horstmanshoff, Helen King, and Claus Zittel (Leiden: Brill Academic, 2012), 503.

1: TO SWEAT IS HUMAN

This chapter benefited greatly from my conversations with many scientists, but particularly Andrew Best, Lucia Dacome, Jason Kalimar, Yana Kamberov, Daniel Lieberman, Duncan Mitchell, Michael Stolberg, and Michael Zech.

9   **60-watt light bulb:** Jacques Machaire et al., "Evaluation of the Metabolic Rate Based on the Recording of the Heart Rate," *Industrial Health* 55, no. 3 (May 2017): 219–32, https://doi.org/10.2486/indhealth.2016-0177.

9   **100-watt light bulb:** Machaire et al., "Evaluation of the Metabolic Rate."

9   **heatstroke is a miserable way to go:** A. Bouchama and J. P. Knochel, "Heat
    Stroke," *New England Journal of Medicine* 346, no. 25 (June 20, 2002): 1978,
    https://pubmed.ncbi.nlm.nih.gov/12075060/. For a nail-biting description,
    see: Amy Ragsdale and Peter Stark, "What It Feels Like to Die from Heat Stroke,"
    Outside Online, June 18, 2019, https://www.outsideonline.com/2398105/heat
    -stroke-signs-symptoms.

10 **Elephants use their enormous ears:** Terrie M. Williams, "Heat Transfer in
    Elephants: Thermal Partitioning Based on Skin Temperature Profiles," *Jour-
    nal of Zoology* 222, no. 2 (1990): 235–45, https://doi.org/10.1111/j.1469-7998
    .1990.tb05674.x; and Conor L. Myhrvold, Howard A. Stone, and Elie Bou-Zeid,
    "What Is the Use of Elephant Hair?," *PLOS ONE* 7, no. 10 (October 10, 2012),
    https://doi.org/10.1371/journal.pone.0047018.

10 **vultures poop on themselves:** Bernd Heinrich, *Why We Run: A Natural His-
    tory* (New York: HarperCollins, 2009), 102.

11 **35 million years ago:** Sweat glands moved from being localized in paws to
    being distributed throughout the entire body "in the common ancestor of
    catarrhines, the group of primates that includes Old World monkeys," as
    described in Yana G. Kamberov et al., "Comparative Evidence for the Inde-
    pendent Evolution of Hair and Sweat Gland Traits in Primates," *Journal of
    Human Evolution* 125 (December 1, 2018): 99–105, https://doi.org/10.1016/j
    .jhevol.2018.10.008. The split of Old World monkeys (catarrhines) to New
    World monkeys (platyrrhines) is thus the point at which our predecessors
    began evolving sweat glands, but when did *that* happen? This is debated, but I
    have gone with 35 million years ago as found by Carlos G. Schrago and Claudia
    A. M. Russo, "Timing the Origin of New World Monkeys," *Molecular Biology
    and Evolution* 20, no. 10 (October 1, 2003): 1620–25, https://doi.org/10.1093/
    molbev/msg172. By 2 million to 3 million years ago, at the start of the *Homo*
    genus, there was a huge expansion in the density of sweat glands.

11 **Sweating allowed us to forage out in the sun:** Provided that we could carry
    water, which was facilitated by being bipedal.

11 **7% of our body's surface area to the intense heat radiating down:** Daniel E.
    Lieberman, "Human Locomotion and Heat Loss: An Evolutionary Perspec-
    tive," *Comprehensive Physiology* 5, no. 1 (January 1, 2015): 99–117, https://doi
    .org/10.1002/cphy.c140011.

12 **2 million and 5 million sweat pores:** Some references quote the range as
    1.6 million to 5 million but most list the range as 2 million to 5 million, in
    reference to an average calculated here: Yas Kuno, *Human Perspiration*
    (Springfield, IL: Charles C. Thomas, 1956), 66. First published in 1934 as *The
    Physiology of Human Perspiration* by J. & A. Churchill, London.

12   **water traversing Niagara Falls on a hot summer day:** According to the Niagara Falls State Park frequently asked questions, "75,750 gallons of water per second over the American and Bridal Veil Falls and 681,750 gallons per second over the Horseshoe Falls," which totals to 757,500 gallons per second over both falls. This converts to 172 million liters per minute flowing over Niagara Falls writ large. Meanwhile, the human sweaters: Let's consider first the extremely sweaty extreme. The flux of a very heavy sweater is 20 nanoliters per minute per pore. And the maximum number of sweat glands is 5 million per person. If all 8 billion humans on Earth were at the high end for flux and pore count, we'd produce a combined flux of 800 million liters of sweat per minute, more than four times what is going over Niagara Falls. But we are all not extremely heavy sweaters. Humans at the other end of the spectrum have a sweat flux of 2 nanoliters per minute per pore, and 2 million sweat pores. This amounts to a combined global flux of 32 million liters of sweat per minute, which is a fifth of the Niagara Falls rate. An average sweater would be somewhere in between.

12   **lose as much as 25 grams of salt:** Graham P. Bates and Veronica S. Miller, "Sweat Rate and Sodium Loss During Work in the Heat," *Journal of Occupational Medicine and Toxicology (London, England)* 3 (January 29, 2008): 4, https://doi.org/10.1186/1745-6673-3-4.

12   **most of us lose only a fraction of this on a day-to-day basis:** Duncan Mitchell, email correspondence with author, July 14, 2020. Mitchell would estimate most people max out at about 12–15 grams of salt per day.

12   **apocrine glands, which become active at puberty:** Catherine Lu and Elaine Fuchs, "Sweat Gland Progenitors in Development, Homeostasis, and Wound Repair," *Cold Spring Harbor Perspectives in Medicine* 4, no. 2 (February 2014), https://doi.org/10.1101/cshperspect.a015222.

13   **Apocrine glands are much bigger than eccrine glands:** Morgan B. Murphrey and Tanvi Vaidya, "Histology, Apocrine Gland," in *StatPearls* (Treasure Island, FL: StatPearls Publishing, 2020), http://www.ncbi.nlm.nih.gov/books/NBK482199/.

13   **Nicotine, cocaine, garlic odor, food dyes, amphetamines, antibiotics:** Melanie J. Bailey et al., "Chemical Characterization of Latent Fingerprints by Matrix-Assisted Laser Desorption Ionization, Time-of-Flight Secondary Ion Mass Spectrometry, Mega Electron Volt Secondary Mass Spectrometry, Gas Chromatography/Mass Spectrometry, X-Ray Photoelectron Spectroscopy, and Attenuated Total Reflection Fourier Transform Infrared Spectroscopic Imaging: An Intercomparison," *Analytical Chemistry* 84, no. 20 (October 16, 2012): 8514–23, https://doi.org/10.1021/ac302441y.

13   **cranberry juice turned his sweat red:** Ashok Kumar Jaiswal, Shilpashree P. Ravikiran, and Prasoon Kumar Roy, "Red Eccrine Chromhidrosis with

Review of Literature," *Indian Journal of Dermatology* 62, no. 6 (2017): 675, https://doi.org/10.4103/ijd.IJD_755_16.

13   **swallowed so many laxatives that he turned his sweat yellow:** Bharathi Sundaramoorthy, "Bisacodyl Induced Chromhidrosis—A Case Report," *University Journal of Medicine and Medical Specialities* 3, no. 4 (July 13, 2017), http://ejournal-tnmgrmu.ac.in/index.php/medicine/article/view/4614.

13   **On occasion our bodies also construct hybrid versions of eccrine and apocrine pores, called apoeccrine glands:** K. Sato and F. Sato, "Sweat Secretion by Human Axillary Apoeccrine Sweat Gland in Vitro," pt. 2, *American Journal of Physiology* 252, no. 1 (January 1987): R181–87, https://doi.org/10.1152/ajpregu.1987.252.1.R181.

14   **waste products from exercise, such as lactic acid and urea, as well as glucose and some metals:** Amay J. Bandodkar et al., "Wearable Sensors for Biochemical Sweat Analysis," *Annual Review of Analytical Chemistry* 12, no. 1 (June 12, 2019): 1–22, https://doi.org/10.1146/annurev-anchem-061318-114910.

14   **proteins that do crowd control:** B. Schittek et al., "Dermcidin: A Novel Human Antibiotic Peptide Secreted by Sweat Glands," *Nature Immunology* 2, no. 12 (December 2001), https://doi.org/10.1038/ni732.

14   **perspiration even carries markers of disease:** Simona Francese, R. Bradshaw, and N. Denison, "An Update on MALDI Mass Spectrometry Based Technology for the Analysis of Fingermarks—Stepping into Operational Deployment," *Analyst* 142, no. 14 (2017): 2518–46, https://doi.org/10.1039/C7AN00569E.

14   **Zech wondered how long it would take for a sip of water to transit from his lips to his sweat pores:** Michael Zech et al., "Sauna, Sweat and Science II—Do We Sweat What We Drink?," *Isotopes in Environmental and Health Studies* 55, no. 4 (July 4, 2019): 394–403, https://doi.org/10.1080/10256016.2019.1635125.

14   **a chemical tracer:** Technically it was an isotope tracer, namely a deuterated compound. But given that a heavy molecule is still a molecule, I've taken creative license and called it a chemical tracer to avoid a treatise that explains isotope analysis to the layperson.

15   **it took less than 15 minutes for the tracer to transit through his stomach:** Zech et al., "Sauna, Sweat and Science II."

15   **Galen proposed that insensible vapors were being discharged continually from the body:** Michael Stolberg, "Sweat. Learned Concepts and Popular Perceptions, 1500–1800," in *Blood, Sweat and Tears: The Changing Concepts of Physiology from Antiquity into Early Modern Europe*, ed. Manfred Horstmanshoff, Helen King, and Claus Zittel (Leiden: Brill Academic, 2012).

16   **"cleansed the body of superfluities and of potentially harmful, dangerous, polluting matter":** Stolberg, "Sweat," 511.

17   **a new dawn of perspiration exploration arose thanks to Santorio Santorio:** Lucia Dacome, "Balancing Acts: Picturing Perspiration in the Long Eighteenth Century," *Studies in History and Philosophy of Science Part C: Studies in History and Philosophy of Biological and Biomedical Sciences* 43, no. 2 (June 2012): 379–91, https://doi.org/10.1016/j.shpsc.2011.10.030.

17   **device for measuring pulse rates:** Fabrizio Bigotti and David Taylor, "The Pulsilogium of Santorio: New Light on Technology and Measurement in Early Modern Medicine," *Societate Si Politica* 11, no. 2 (2017): 53–113, https://www.ncbi.nlm.nih.gov/pmc/articles/PMC6407692/. And also: Richard de Grijs and Daniel Vuillermin, "Measure of the Heart: Santorio Santorio and the Pulsilogium," arXiv:1702.05211 [physics.hist-ph], February 17, 2017, http://arxiv.org/abs/1702.05211.

17   **Santorio devised an elaborate hanging chair:** Dacome, "Balancing Acts."

17   **Jan Evangelista Purkyně discovered sweat's exit portal:** "Jan Evangelista Purkinje | Czech Physiologist," in *Encyclopedia Britannica*, accessed July 16, 2020, https://www.britannica.com/biography/Jan-Evangelista-Purkinje.

17   **Swiss and German physiologists recorded the electrical signals:** I'm referring to Hermann and Luchsinger in 1878 as referenced in Wolfram Boucsein, *Electrodermal Activity* (New York: Springer Science & Business Media, 2012), 4.

18   **a whole-body technique for visualizing the onset of sweat:** Victor Minor, "Ein neues Verfahren zu der klinischen Untersuchung der Schweißabsonderung," *Deutsche Zeitschrift für Nervenheilkunde* 101, no. 1 (January 1, 1928): 302–8, https://doi.org/10.1007/BF01652699.

18   **the Achilles tendon, for example, or the bald heads of men:** Minor, "Ein neues Verfahren."

19   **They also measured the resistance of different layers of skin by *thrusting*:** Kuno, *Human Perspiration*, 9.

19   **about 70 micrometers:** Kuno, *Human Perspiration*, 18.

19   **2 million and 5 million eccrine pores:** Kuno, *Human Perspiration*, 66. Some modern-day reports have extended the lower end of the sweat gland count to 1.6 million pores.

19   **famous 1934 tome:** Taketoshi Morimoto, "History of Thermal and Environmental Physiology in Japan: The Heritage of Yas Kuno," *Temperature: Multidisciplinary Biomedical Journal* 2, no. 3 (July 28, 2015): 310–15, https://doi.org/10.1080/23328940.2015.1066920. (Note I have only been able to read the 1956 reprint of Kuno's work referenced earlier.)

19   **was hired to figure out how much water to supply the infantry:** Many scientists within the military have also studied and continue to study sweat, such as those at the US Army Natick Soldier Systems Center. My requests to interview them and visit the center languished.

20 **"Various methods have been tried . . . his attention":** Edward Frederick Adolph, *Physiology of Man in the Desert* (New York: Interscience, 1947), 10.

20 **were only instituted by Congress in 1973:** "Research Ethics Timeline," National Institute of Environmental Health Sciences, accessed July 22, 2020, https://www.niehs.nih.gov/research/resources/bioethics/timeline/index.cfm.

20 **"become emotionally unstable":** Adolph, *Physiology of Man*, 14.

20 **"Only desire to stop and rest":** Rick Lovett, "The Man Who Revealed the Secrets of Sweat," *New Scientist*, accessed July 15, 2020, https://www.newscientist.com/article/mg20227061-500-the-man-who-revealed-the-secrets-of-sweat/.

20 **"Unsociable attitude . . . finally stopped":** Adolph, *Physiology of Man*, 214.

20 **variables that the US Army now inputs into fancy computer algorithms to estimate the water needs of soldiers:** Richard R. Gonzalez et al., "Sweat Rate Prediction Equations for Outdoor Exercise with Transient Solar Radiation," *Journal of Applied Physiology* 112, no. 8 (April 15, 2012): 1300–10, https://doi.org/10.1152/japplphysiol.01056.2011.

21 **"A soldier walking at the rate of 3.5 miles . . . 1 cup of sweat":** Adolph, *Physiology of Man*, 4.

21 **"It is better to have the water inside you than to carry it":** Lovett, "Secrets of Sweat."

21 **did PhD research:** Schneider told me that to do this research, she and her subjects would drink water containing radioactive tracers.

21 **there's little evidence:** Lindsay B. Baker, "Physiology of Sweat Gland Function: The Roles of Sweating and Sweat Composition in Human Health," *Temperature* 6, no. 3 (July 3, 2019): 211–59, https://doi.org/10.1080/23328940.2019.1632145.

22 **long pants and long-sleeved T-shirts in extremely hot and dry environments:** Lovett, "Secrets of Sweat."

22 **primarily Black men working for extremely little pay:** Black men were assigned to manual labor in the mines while white men served in supervisory positions, according to many sources, including Suzanne M. Schneider, "Heat Acclimation: Gold Mines and Genes," *Temperature: Multidisciplinary Biomedical Journal* 3, no. 4 (September 27, 2016): 527–38, https://doi.org/10.1080/23328940.2016.1240749. Also: "The workforce consisted of a unionised white labour aristocracy employed in supervisory roles, while the bulk of the manual work was done by black migrants." Jock McCulloch, *South Africa's Gold Mines and the Politics of Silicosis* (Woodbridge, UK: James Currey, 2012).

22 **Within a few decades, miners needed to go half a mile underground to extract it:** Schneider, "Heat Acclimation."

23 **they were descending nearly 2 miles below:** Schneider, "Heat Acclimation."

23 **"working in the grave":** Matthew John Smith, " 'Working in the Grave': The Development of a Health and Safety System on the Witwatersrand Gold Mines 1900–1939" (master's thesis, Rhodes University, 1993).

23 **most miners entered an elevator that could hold 100-plus people:** Miners are still descending in this way, as described by Matthew Hart, "A Journey Into the World's Deepest Gold Mine," *Wall Street Journal*, December 13, 2013, Personal Finance, https://www.wsj.com/articles/a-journey-into-the-world8217s -deepest-gold-mine-1386951413; and Duncan Mitchell, telephone interview by the author, August 28, 2018.

23 **at full speed:** Hart, "A Journey"; Mitchell, telephone interview with author.

23 **In more recent times, accidents involving gold mine elevators in South Africa killed 105 miners in 1995 and 9 miners in 2009:** Hart, "A Journey."

23 **The rock didn't just hold gold; it also contained silica, which, when blasted apart, produces dust that is extremely harmful to lungs:** Jock McCulloch, "Dust, Disease and Politics on South Africa's Gold Mines," *Adler Museum Bulletin*, June 2013, https://www.wits.ac.za/media/migration/files/cs-38933 -fix/migrated-pdf/pdfs-4/Adler%20Bulletin%20June%202013.pdf.

23 **"The earth will swallow us who burrow . . . all round me, every day, I see men stumble, fall and die":** Benedict Wallet Vilakazi wrote the poem "Ezinkomponi," or "In the Gold Mines," published in *Zulu Horizons* (Johannesburg: Wits University Press, 1973).

24 **class action suits from 500,000 miners:** "South Africa Allows Silicosis Class Action Against Gold Firms," Reuters, May 13, 2016, https://www.reuters .com/article/us-safrica-gold-silicosis-idUSKCN0Y40Q2.

24 **Hundreds of Witwatersrand gold miners died of heatstroke:** Getting an exact number has been a challenge. Looking at data discussed in M. J. Martinson, "Heat Stress in Witwatersrand Gold Mines," *Journal of Occupational Accidents* 1, no. 2 (January 1, 1977): 171–93, https://doi.org/10.1016/0376 -6349(77)90013-X, Schneider, "Heat Acclimation," and Smith, " 'Working in the Grave,' " the first Witwatersrand heatstroke deaths were reported in 1924, and by 1931 there were more than 92 additional heatstroke deaths. From 1956 to 1961 there were 47 fatalities. Using these as well as other heatstroke fatalities, I estimate that there were approximately 10 heatstroke deaths per year. In the ~40 years between 1924 and the mid-1960s (when new acclimation measures to prevent heatstroke were instituted), that would amount to about 400 deaths.

24 **hiring or funding medical researchers:** Among these medical researchers were Aldo Dreosti, who worked for Rand Mines Ltd from 1927 to 1963, and Cyril Wyndham, who worked for the Human Science Laboratory at the Chamber of Mines in Johannesburg. Wyndham's papers with colleague N. B. Stry-

dom are still referenced by heat-acclimation researchers today. For a more thorough outline of the research performed at the Human Science Laboratory, read: Cyril H. Wyndham, "Adaptation to Heat and Cold," *Environmental Research* 2 (1969): 442–69.

24   **Deaths by heatstroke were reduced—although not entirely eliminated:** Martinson, "Heat Stress in Witwatersrand Gold Mines." Martinson notes in Table 5 that there were 20 heatstroke deaths between 1968 and 1973 in South African gold mines, which is about an average of 3 to 4 deaths per year, down from ~10 heatstroke deaths in gold mines per year.

25   **By the mid-1960s:** Schneider, "Heat Acclimation"; Mitchell, telephone interview with author.

25   **4 hours a day in a hot, humid tent doing step-up exercises on blocks, with periodic rectal temperature checks:** Schneider, "Heat Acclimation."

25   **widely disliked:** Schneider, "Heat Acclimation."

25   **vests that contained dry ice:** Bill Whitaker, "What Lies at the Bottom of One of the Deepest Holes Ever Dug by Man?," *60 Minutes*, CBS News, August 4, 2019, https:// www.cbsnews.com/news/south-africa-gold-mining-what-lies-at-the-bottom-of -one-of-the-deepest-holes-ever-dug-by-man-60-minutes-2019-08-04/.

25   **athletes complete similarly structured:** Michael N. Sawka, "Heat Acclimatization to Improve Athletic Performance in Warm-Hot Environments," *Sports Science Exchange* 28, no. 153 (2015): 7. And Sébastien Racinais et al., "Heat Acclimation," in *Heat Stress in Sport and Exercise—Thermophysiology of Health and Performance*, ed. Julien D. Périard and Sébastien Racinais (Basel: Springer International, 2019), https://doi.org/10.1007/978-3-319 -93515-7.

25   **60 to 90 minutes at a time:** Racinais et al., "Heat Acclimation."

25   **anatomy, infectious disease, and psychiatry:** I would encourage anyone who doubts this statement to have a close look at the National Institutes of Health's sobering timeline of ethical issues in human research: "Research Ethics Timeline," National Institute of Environmental Health Sciences, accessed July 22, 2020, https://www.niehs.nih.gov/research/resources/bioethics/ timeline/index.cfm. I would also encourage anyone who has not already done so to immediately read Rebecca Skloot, *The Immortal Life of Henrietta Lacks* (New York: Broadway Books, 2011). First published 2010 by Crown, New York.

25   **"commenced with the inauguration of a colonial policy by the European nations":** E. S. Sundstroem, "The Physiological Effects of Tropical Climate," *Physiological Reviews* 7, no. 2 (April 1, 1927): 320–62, https://doi.org/10.1152/ physrev.1927.7.2.320.

25   **"the chances of white settlement in the tropics":** Sundstroem, "Physiological Effects."

26   "sweating is the only process which makes human life comfortable during hot weather and therefore human existence possible in the torrid zone": Kuno, *Human Perspiration*, 3.

29   "If we look at the shadow of a bare head . . . mounting upward": Jacobus Benignus Winsløw, *An Anatomical Exposition of the Structure of the Human Body*, 4th ed., corrected, trans. G. Douglas (London: R. Ware, J. Knapton, S. Birt, T. and T. Longman, C. Hitch and L. Hawes, C. Davis, T. Astley, and R. Baldwin, 1756).

29   "Lord, I was tired. My heart was motoring at about 155 beats per minute, I was sweating like a pig, the pickle was a pest, and I had yet to begin my real work": Eugene Cernan and Donald A. Davis, *The Last Man on the Moon: Astronaut Eugene Cernan and America's Race in Space* (New York: St. Martin's, 2007).

29   others think he was referring to his penis: "Idioms, What Do They Mean?," The Cellar, accessed July 15, 2020, https://cellar.org/showthread .php?t=23318&page=3.

29   the 13 pounds of sweat Cernan lost: "Gene Cernan, Last Astronaut on the Moon, Dies at 82," *Tribune News Services*, accessed July 15, 2020, https://www .chicagotribune.com/nation-world/ct-gene-cernan-dead-20170116-story.html.

30   "I was as weary as I had ever been in my life": Cernan and Davis, *The Last Man on the Moon*.

2: SWEAT LIKE A PIG

This chapter was by far the most fun and delightfully gross chapter to research and write. I am grateful for conversations with Kathrin Dausmann, Yana Kamberov, Danielle Levesque, Duncan Mitchell, and Blair Wolf.

31   **Roger Gentry measured the copulation frequency:** Roger L. Gentry, "Thermoregulatory Behavior of Eared Seals," *Behaviour* 46, no. 1/2 (1973): 73–93, https://www.jstor.org/stable/4533520.

31   **"At high temperatures, land-locked male seals . . . rear flippers. They then lay on one side and extended the wet rear flipper into the air":** Gentry, "Thermoregulatory Behavior of Eared Seals."

32   **"regurgitate their stomach contents . . . their forefeet":** Bernd Heinrich, *Why We Run: A Natural History* (New York: HarperCollins, 2009).

32   **"lick[ing] off the residual solids that are left after the water has evaporated":** Heinrich, *Why We Run*.

32   **"a turkey vulture sitting on a fence . . . makes sense":** Heinrich, *Why We Run*.

32 **Water evaporation...most efficient strategy for cooling down in hot weather:** Duncan Mitchell et al., "Revisiting Concepts of Thermal Physiology: Predicting Responses of Mammals to Climate Change," in "Linking Organismal Functions, Life History Strategies and Population Performance," ed. Dehua Wang, special feature, *Journal of Animal Ecology* 87, no. 4 (July 2018): 956–73, https://doi.org/10.1111/1365-2656.12818.

33 **A giraffe is a classic dolichomorphic animal:** Mitchell et al., "Revisiting Concepts of Thermal Physiology."

33 **avoid the intense heat of the noonday sun by changing the color of their skin:** Kathleen R. Smith et al., "Colour Change on Different Body Regions Provides Thermal and Signalling Advantages in Bearded Dragon Lizards," *Proceedings of the Royal Society B: Biological Sciences* 283, no. 1832 (June 15, 2016), https://doi.org/10.1098/rspb.2016.0626. And Kathleen R. Smith et al., "Color Change for Thermoregulation versus Camouflage in Free-Ranging Lizards," *American Naturalist* 188, no. 6 (December 2016), https://doi.org/10.1086/688765.

34 **an elephant's body is glowing hot while its ears are not:** Mitchell et al., "Revisiting Concepts of Thermal Physiology."

34 **a torpor state:** Julia Nowack et al., "Variable Climates Lead to Varying Phenotypes: 'Weird' Mammalian Torpor and Lessons from Non-Holarctic Species," *Frontiers in Ecology and Evolution* 8 (2020), https://doi.org/10.3389/fevo.2020.00060.

34 **space scientists keen to find out whether humans might also be able to torpor:** John Bradford, "Torpor Inducing Transfer Habitat for Human Stasis to Mars," NASA, August 7, 2017, http://www.nasa.gov/content/torpor-inducing-transfer-habitat-for-human-stasis-to-mars.

34 **Pigs wallow in mud to keep cool:** Edith J. Mayorga et al., "Heat Stress Adaptations in Pigs," *Animal Frontiers* 9, no. 1 (January 3, 2019): 54–61, https://doi.org/10.1093/af/vfy035.

35 **koala:** Natalie J. Briscoe et al., "Tree-Hugging Koalas Demonstrate a Novel Thermoregulatory Mechanism for Arboreal Mammals," *Biology Letters* 10, no. 6 (June 2014), https://doi.org/10.1098/rsbl.2014.0235.

35 **kangaroos is to lick their forearms:** Terence J. Dawson et al., "Thermoregulation by Kangaroos from Mesic and Arid Habitats: Influence of Temperature on Routes of Heat Loss in Eastern Grey Kangaroos (*Macropus giganteus*) and Red Kangaroos (*Macropus rufus*)," *Physiological and Biochemical Zoology* 73, no. 3 (May 2000): 374–81, https://doi.org/10.1086/316751.

38 **Gular flutter:** Eric Krabbe Smith et al., "Avian Thermoregulation in the Heat: Resting Metabolism, Evaporative Cooling and Heat Tolerance in Sonoran Desert Doves and Quail," *Journal of Experimental Biology* 218, no. 22 (November

1, 2015): 3636–46, https://doi.org/10.1242/jeb.128645. It's worth pointing out that gular flutter is also used by birds that don't have pouches.

38 **The tawny frogmouth, . . . 100 breaths per minute at 108.5°F:** Robert C. Lasiewski and George A. Bartholomew, "Evaporative Cooling in the Poor-Will and the Tawny Frogmouth," *Condor* 68, no. 3 (1966): 253–62, https://doi .org/10.2307/1365559. Note the paper refers to this rate when the bird's body temperature rises to 42.5°C (which I have converted to 108.5°F).

38 **eggs begin to cook at about 104°F, desert doves have to constantly cool their bodies:** Glenn E. Walsberg and Katherine A. Voss-Roberts, "Incubation in Desert-Nesting Doves: Mechanisms for Egg Cooling," *Physiological Zoology* 56, no. 1 (1983): 88–93, http://www.jstor.org/stable/30159969.

38 **One of my favorite studies about a pigeon's ability to air-condition its eggs:** Walsberg and Voss-Roberts, "Incubation in Desert-Nesting Doves."

39 **where it can get as hot as 120°F:** During the time of the particular experiment described in the paper, the desert temperature was at 113°F, but Walsberg says 120°F is common.

39 **"Before measuring body temperature . . . 22 attempts":** Walsberg and Voss-Roberts, "Incubation in Desert-Nesting Doves."

39 **whopping 25°F below ambient air temperature:** Krabbe Smith et al., "Avian Thermoregulation in the Heat."

40 **then flushing the fluid out pores in its abdomen:** Neil F. Hadley, Michael C. Quinlan, and Michael L Kennedy, "Evaporative Cooling in the Desert Cicada: Thermal Efficiency and Water/Metabolic Cost," *Journal of Experimental Biology* 159 (1991): 269–83.

40 **multiple species of frogs:** William A. Buttemer, "Effect of Temperature on Evaporative Water Loss of the Australian Tree Frogs *Litoria caerulea* and *Litoria chloris*," *Physiological Zoology* 63, no. 5 (1990): 1043–57, https://www .jstor.org/stable/30152628; and Mohlamatsane Mokhatla, John Measey, and Ben Smit, "The Role of Ambient Temperature and Body Mass on Body Temperature, Standard Metabolic Rate and Evaporative Water Loss in Southern African Anurans of Different Habitat Specialisation," *PeerJ* 7 (2019), https:// doi.org/10.7717/peerj.7885.

42 **"ten times the density of eccrine glands":** Yana G. Kamberov et al., "Comparative Evidence for the Independent Evolution of Hair and Sweat Gland Traits in Primates," *Journal of Human Evolution* 125 (December 1, 2018): 99–105, https://doi.org/10.1016/j.jhevol.2018.10.008.

42 **began searching for an answer:** Yana G. Kamberov et al., "A Genetic Basis of Variation in Eccrine Sweat Gland and Hair Follicle Density," *Proceedings of the National Academy of Sciences USA*, July 16, 2015, https://doi.org/10.1073/ pnas.1511680112.

42 **Horses also sweat:** C. M. Scott, D. J. Marlin, and R. C. Schroter, "Quantifica-
tion of the Response of Equine Apocrine Sweat Glands to Beta2-Adrenergic
Stimulation," *Equine Veterinary Journal* 33, no. 6 (November 2001): 605–12,
https://doi.org/10.2746/042516401776563463.

43 **sweaty horses at a start line is a bad omen:** "Physiology of a Thorough-
bred," Blinkers On Racing Stable blog, accessed July 16, 2020, https://
blinkersonracing.wordpress.com/category/physiology-of-a-thoroughbred/.

43 **"sweating on its own was not a reliable performance indicator, but in conjunction with other variables might indicate losers":** G. D. Hutson and
M. J. Haskell, "Pre-Race Behaviour of Horses as a Predictor of Race Finish-
ing Order," *Applied Animal Behaviour Science* 53, no. 4 (July 1, 1997): 231–48,
https://doi.org/10.1016/S0168-1591(96)01162-8.

43 **Consider a cow:** An impressive number of researchers have measured sweat
rates in cows, using a cornucopia of devices including my favorite: "the bovine
evaporation meter." K. G. Gebremedhin et al., "Sweating Rates of Dairy Cows
and Beef Heifers in Hot Conditions," *Trans ASABE* 51 (January 1, 2008),
https://doi.org/10.13031/2013.25397. Some researchers have compared sweat
rates on black versus white cow skin: Roberto Gomes da Silva and Alex Sandro
Campos Maia, "Evaporative Cooling and Cutaneous Surface Temperature of
Holstein Cows in Tropical Conditions," *Revista Brasileira de Zootecnia* 40, no.
5 (May 2011): 1143–47, https://doi.org/10.1590/S1516-35982011000500028.
And researchers in Portugal even developed a new device for improved accuracy of sweat rate measurements, which involves gluing Velcro onto shaved cow
skin: Alfredo Manuel Franco Pereira et al., "A Device to Improve the Schleger
and Turner Method for Sweating Rate Measurements," *International Journal
of Biometeorology* 54, no. 1 (January 1, 2010): 37–43, https://doi.org/10.1007/
s00484-009-0250-3. Suffice to say that I opted for a high bovine sweat rate of
150 g/m² h for this back-of-the-envelope calculation.

43 **a heavily sweating human:** For this back-of-the-envelope calculation, I esti-
mated that a heavily sweating human produces 2 liters of sweat per hour,
which converts to 0.033 liter per minute, which converts to 6.75 teaspoons per
minute.

44 **a reddish-pink sweat that acts as a sunscreen:** Yoko Saikawa et al., "The Red
Sweat of the Hippopotamus," *Nature* 429, no. 6990 (May 2004): 363, https://
doi.org/10.1038/429363a.

44 **a cooling parasol:** See this paper for a very cool thermal image of the cooling
hump: Khalid A. Abdoun et al., "Regional and Circadian Variations of Sweating Rate and Body Surface Temperature in Camels (*Camelus dromedarius*),"
Animal Science Journal 83, no. 7 (2012): 556–61, https://doi.org/10.1111/j
.1740-0929.2011.00993.x.

44 **a whopping 10°F:** (Or 6°C) Knut Schmidt-Nielsen et al., "Body Temperature of the Camel and Its Relation to Water Economy," *American Journal of
Physiology-Legacy Content* 188, no. 1 (December 31, 1956): 103–12, https://
doi.org/10.1152/ajplegacy.1956.188.1.103. And: Hanan Bouâouda et al., "Daily
Regulation of Body Temperature Rhythm in the Camel (*Camelus dromedarius*) Exposed to Experimental Desert Conditions," *Physiological Reports* 2,
no. 9 (September 2014), https://doi.org/10.14814/phy2.12151.

44 **Water evaporating off the moist membranes of the camel's nasal cavity
cools nearby blood traversing up into the brain:** A. O. Elkhawad, "Selective
Brain Cooling in Desert Animals: The Camel (*Camelus dromedarius*)," *Comparative Biochemistry and Physiology Part A: Physiology* 101, no. 2 (February
1992), https://doi.org/10.1016/0300-9629(92)90522-r.

44 **"a temperature-sensitive sphincter" to specifically divert cool nasal blood
to the brain during heat stress:** A. O. Elkhawad, N. S. Al-Zaid, and M. N.
Bou-Resli, "Facial Vessels of Desert Camel (*Camelus dromedarius*): Role in
Brain Cooling," *American Journal of Physiology—Regulatory, Integrative
and Comparative Physiology* 258, no. 3 (March 1, 1990): R602–7, https://doi
.org/10.1152/ajpregu.1990.258.3.R602.

3: THE SWEET SMELL OF YOU

This chapter was made possible thanks to illuminating conversations with Chris
Callewaert, Pamela Dalton, Johan Lundström, Mats Olsson, George Preti, and Annlyse Retiveau (who additionally was kind enough to smell my armpits).

48 **particularly a genus called *Corynebacterium*:** A. Gordon James et al.,
"Microbiological and Biochemical Origins of Human Axillary Odour," *FEMS
Microbiology Ecology* 83, no. 3 (March 1, 2013): 527–40, https://doi.org/10
.1111/1574-6941.12054.

48 **Most deodorants contain antiseptics:** Karl Laden, *Antiperspirants and
Deodorants,* 2nd ed. (Boca Raton, FL: CRC Press, 1999).

48 **feature fragrances as backup:** Laden, *Antiperspirants and Deodorants.*

48 **sensory analysts sniff one pit, pause to let their nose clear, and then slide
over to the other pit for comparison:** ASTM E1207-09, *Standard Guide for
Sensory Evaluation of Axillary Deodorancy* (West Conshohocken, PA: ASTM
International, February 1, 2009), https://doi.org/10.1520/E1207-09.

49 ***Standard Guide for Sensory Evaluation of Axillary Deodorancy:*** ASTM
E1207-09, *Standard Guide.*

50 **"pre-laundered wearing apparel":** ASTM E1207-09, *Standard Guide.*

50 **"Axillae should only be . . . bathing or showering at home":** ASTM E1207-09, *Standard Guide.*

52 **The body-odor wheel's:** "The B.O. Wheel," *Slate*, March 25, 2009, https://slate .com/technology/2009/03/the-b-o-wheel.html.

53 **Stasi spy agency collected sweat samples of dissidents and other enemies of the state:** Thomas Darnstädt et al., "Stasi Methods Used to Track G8 Opponents: The Scent of Terror," *Der Spiegel—International*, May 23, 2007, https://www.spiegel.de/international/germany/stasi-methods-used-to-track -g8-opponents-the-scent-of-terror-a-484561.html.

53 **A West German man was convicted of murder in 1989:** Darnstädt et al., "Stasi Methods."

54 **left-wing activists whom they suspected might disrupt a G8 meeting:** Darnstädt et al., "Stasi Methods."

54 **more *Corynebacterium* present in a person's armpit microbiome means that more sulfurous, stinky molecules waft off the armpit:** James et al., "Microbiological and Biochemical Origins."

54 ***Anaerococcus*:** Takayoshi Fujii et al., "A Newly Discovered *Anaerococcus* Strain Responsible for Axillary Odor and a New Axillary Odor Inhibitor, Pentagalloyl Glucose," *FEMS Microbiology Ecology* 89, no. 1 (July 2014): 198– 207, https://doi.org/10.1111/1574-6941.12347.

54 ***Micrococcus*:** James et al., "Microbiological and Biochemical Origins."

55 **In 1992, Preti and his colleagues discovered that the top note:** Xiao-Nong et al., "An Investigation of Human Apocrine Gland Secretion for Axillary Odor Precursors," *Journal of Chemical Ecology* 18 (July 1992): 1039–55, https://doi .org/10.1007/BF00980061.

55 **In a study done by the Swiss company Firmenich:** M. Troccaz et al., "Gender-Specific Differences between the Concentrations of Nonvolatile (R)/(S)-3-Methyl-3-Sulfanylhexan-1-Ol and (R)/(S)-3-Hydroxy-3-Methyl-Hexanoic Acid Odor Precursors in Axillary Secretions," *Chemical Senses* 34, no. 3 (December 16, 2008): 203–10, https://doi.org/10.1093/chemse/bjn076.

58 **800 genes in the mammalian genome that code for odor receptors:** Tsviya Olender, Doron Lancet, and Daniel W. Nebert, "Update on the Olfactory Receptor (OR) Gene Superfamily," *Human Genomics* 3, no. 1 (September 1, 2008): 87–97, https://doi.org/10.1186/1479-7364-3-1-87.

59 **to detect the airborne molecules that Milne can smell:** Drupad K. Trivedi et al., "Discovery of Volatile Biomarkers of Parkinson's Disease from Sebum," *ACS Central Science* 5, no. 4 (April 24, 2019): 599, https://doi.org/10.1021/ acscentsci.8b00879.

59 **ovarian cancer:** Lorenzo Ramirez et al., "Exploring Ovarian Cancer Detec-

tion Using an Interdisciplinary Investigation of Its Volatile Odor Signature," *Journal of Clinical Oncology* 36, no. 15 suppl (May 20, 2018): e17524–e17524, https://doi.org/10.1200/JCO.2018.36.15_suppl.e17524.

59 **smell when another person's immune system is activated from a pathogen:** Mats J. Olsson et al., "The Scent of Disease: Human Body Odor Contains an Early Chemosensory Cue of Sickness," *Psychological Science* 25, no. 3 (January 22, 2014): 817–23, https://doi.org/10.1177/0956797613515681.

61 **plenty of evidence that humans produce an identifiable odor when we feel fear or anxiety:** Jasper H. B. de Groot, Monique A. M. Smeets, and Gün R. Semin, "Rapid Stress System Drives Chemical Transfer of Fear from Sender to Receiver," *PLOS ONE* 10, no. 2 (February 27, 2015), https://doi.org/10.1371/journal.pone.0118211; and Pamela Dalton et al., "Chemosignals of Stress Influence Social Judgments," *PLOS ONE* 8, no. 10 (October 9, 2013): e77144, https://doi.org/10.1371/journal.pone.0077144.

61 **so stressed out that they get sweaty:** Annette Martin et al., "Effective Prevention of Stress-Induced Sweating and Axillary Malodour Formation in Teenagers," *International Journal of Cosmetic Science* 33 (February 1, 2011): 90–97, https://doi.org/10.1111/j.1468-2494.2010.00596.x.

61 **Trier Social Stress Test:** Johanna U. Frisch, Jan A. Häusser, and Andreas Mojzisch, "The Trier Social Stress Test as a Paradigm to Study How People Respond to Threat in Social Interactions," *Frontiers in Psychology* 6 (February 2, 2015), https://doi.org/10.3389/fpsyg.2015.00014.

62 **collected sweat from donors who were watching either clips of a horror movie or a BBC documentary about Yellowstone National Park:** Jasper H. B. de Groot, Gün R. Semin, and Monique A. M. Smeets, "Chemical Communication of Fear: A Case of Male–Female Asymmetry," *Journal of Experimental Psychology: General* 143, no. 4 (2014): 1515–25, https://doi.org/10.1037/a0035950; and Jasper H. B. de Groot, Gün R. Semin, and Monique A. M. Smeets, "I Can See, Hear, and Smell Your Fear: Comparing Olfactory and Audiovisual Media in Fear Communication," *Journal of Experimental Psychology: General* 143, no. 2 (2014): 825–34, https://doi.org/10.1037/a0033731.

63 **Others have speculated—controversially:** Gérard Brand and Jean-Louis Millot, "Sex Differences in Human Olfaction: Between Evidence and Enigma," *Quarterly Journal of Experimental Psychology B: Comparative and Physiological Psychology* 54B, no. 3 (2001): 259–70, https://doi.org/10.1080/0272499 0143000045.

63 **Tom Mangold's memoir *Splashed!*:** Tom Mangold, *Splashed!: A Life from Print to Panorama* (London: Biteback, 2016).

64 **eyewitness misidentification "is by far . . . erroneous identifications from**

victims or witnesses": Innocence Project, "In Focus: Eyewitness Misidentification," October 21, 2008, https://www.innocenceproject.org/in-focus-eyewitness-misidentification/.

64 **participants were shown a video of a violent crime**: Laura Alho et al., "Nose-witness Identification: Effects of Lineup Size and Retention Interval," *Frontiers in Psychology* 7 (May 30, 2016), https://doi.org/10.3389/fpsyg.2016.00713.

65 **analyzed the body odors of nearly 200 residents of a small Alpine town**: Dustin J. Penn et al., "Individual and Gender Fingerprints in Human Body Odour," *Journal of the Royal Society Interface* 4, no. 13 (April 22, 2007): 331–40, https://doi.org/10.1098/rsif.2006.0182.

65 **people of East Asian descent, it is sprinkled in DNA worldwide**: Koh-ichiro Yoshiura et al., "A SNP in the ABCC11 Gene Is the Determinant of Human Earwax Type," *Nature Genetics* 38, no. 3 (March 2006): 324–30, https://doi.org/10.1038/ng1733.

66 **If you're lucky enough to have two copies of the recessive sequence, your armpits are *relatively* less stinky**: Mark Harker et al., "Functional Characterisation of a SNP in the ABCC11 Allele—Effects on Axillary Skin Metabolism, Odour Generation and Associated Behaviours," *Journal of Dermatological Science* 73, no. 1 (January 2014): 23–30, https://doi.org/10.1016/j.jdermsci.2013.08.016.

66 **Patrick Süskind's novel *Perfume***: Patrick Süskind, *Perfume: The Story of a Murderer* (New York: Vintage, 2001). First published in German in 1985 by Diogenes Verlag, Zurich; first English translation published 1986 by Alfred A. Knopf, New York.

4: LOVE STINKS

I am endlessly thankful to the many people who have spoken to me about their odorous preferences and/or their research on these topics. I am particularly indebted to Tristram Wyatt for his incredible textbook on pheromones and also to George Preti, who was one helluva scientist and a mensch. Rest in Peace. I also very much appreciated conversations with Mats Olsson, Johan Lundström, Bettina Pause, and Claus Wedekind.

71 **"the force of attraction"**: The Polytech Festival of science is organized by Moscow's Polytechnic Museum. "Festival Polytech, 27–28 May 2017, Gorky Park," Polytechnic Museum, http://fest.polymus.ru/en/.

74 **skootch preferentially toward its own birth mother's odor**: H. Varendi and R. H. Porter, "Breast Odour as the Only Maternal Stimulus Elicits Crawling

Towards the Odour Source," *Acta Paediatrica (Oslo, Norway: 1992)* 90, no. 4 (April 2001): 372–75, http://www.ncbi.nlm.nih.gov/pubmed/11332925; and Sébastien Doucet et al., "The Secretion of Areolar (Montgomery's) Glands from Lactating Women Elicits Selective, Unconditional Responses in Neonates," *PLOS ONE* 4, no. 10 (October 23, 2009): e7579, https://doi.org/10.1371/journal.pone.0007579.

74   **a mother can identify her own newborn baby by smell just a few hours after birth. (A parent who didn't give birth can do it too after 72 hours):** Tristram D. Wyatt, *Pheromones and Animal Behavior: Chemical Signals and Signatures*, 2nd ed. (New York: Cambridge University Press, 2014), 279, https://doi.org/10.1017/CBO9781139030748.

74   **odors activated the reward center of the brain:** Johan N. Lundström et al., "Maternal Status Regulates Cortical Responses to the Body Odor of Newborns," *Frontiers in Psychology* 4 (2013), https://doi.org/10.3389/fpsyg.2013.00597.

74   **Siblings and married couples . . . signature mixture of chemicals floating off their bodies:** Wyatt, *Pheromones and Animal Behavior*, 278–79.

75   **anosmia:** Ilona Croy, Viola Bojanowski, and Thomas Hummel, "Men Without a Sense of Smell Exhibit a Strongly Reduced Number of Sexual Relationships, Women Exhibit Reduced Partnership Security—A Reanalysis of Previously Published Data," *Biological Psychology* 92, no. 2 (February 1, 2013): 292–94, https://doi.org/10.1016/j.biopsycho.2012.11.008.

75   **"Which organic sense is the most ungrateful . . . fleeting and transient":** Ann-Sophie Barwich, "A Sense So Rare: Measuring Olfactory Experiences and Making a Case for a Process Perspective on Sensory Perception," *Biological Theory* 9, no. 3 (September 1, 2014): 258–68, https://doi.org/10.1007/s13752-014-0165-z.

75   **"Poor human olfaction is a 19th-century myth":** John P. McGann, "Poor Human Olfaction Is a 19th-Century Myth," *Science* 356, no. 6338 (May 12, 2017), https://doi.org/10.1126/science.aam7263.

76   **"is actually quite large in absolute terms and contains a similar number of neurons to that of other . . . states are influenced by our sense of smell":** McGann, "Poor Human Olfaction."

76   **sniff out a trail of chocolate:** Jess Porter et al., "Mechanisms of Scent-Tracking in Humans," *Nature Neuroscience* 10, no. 1 (January 2007): 27–29, https://doi.org/10.1038/nn1819. Technically the study was *published* in 2007, so the research may have been done in the preceding years.

78   **what people did with their hands after a handshake:** Idan Frumin et al., "A Social Chemosignaling Function for Human Handshaking," *eLife* 4 (March 3, 2015): e05154, https://doi.org/10.7554/eLife.05154.

78 **"only the tip of the iceberg":** Frumin et al., "A Social Chemosignaling Function."

79 **a few components in body odor that appear more intensely in men or in women:** Dustin J. Penn et al., "Individual and Gender Fingerprints in Human Body Odour," *Journal of the Royal Society Interface* 4, no. 13 (April 22, 2007): 331–40, https://doi.org/10.1098/rsif.2006.0182.

79 **asked lesbians, gay men, straight men, and straight women to collect their armpit sweat in cotton pads:** Yolanda Martins et al., "Preference for Human Body Odors Is Influenced by Gender and Sexual Orientation," *Psychological Science* 16, no. 9 (September 2005): 694–701, https://doi.org/10.1111/j.1467-9280.2005.01598.x.

80 **study published by Claus Wedekind in 1995:** Claus Wedekind et al., "MHC-Dependent Mate Preferences in Humans," *Proceedings of the Royal Society B: Biological Sciences* 260, no. 1359 (June 22, 1995): 245–49, https://doi.org/10.1098/rspb.1995.0087.

81 **oral contraceptive pill:** Claus Wedekind and Sandra Füri, "Body Odour Preferences in Men and Women: Do They Aim for Specific MHC Combinations or Simply Heterozygosity?," *Proceedings of the Royal Society B: Biological Sciences* 264, no. 1387 (October 22, 1997): 1471–79, https://doi.org/10.1098/rspb.1997.0204. Also see: Craig Roberts et al., "MHC-Correlated Odour Preferences in Humans and the Use of Oral Contraceptives," *Proceedings of the Royal Society B: Biological Sciences* 275, no. 1652 (December 7, 2008): 2715–22, https://doi.org/10.1098/rspb.2008.0825.

82 **lap dances:** Geoffrey Miller, Joshua M. Tybur, and Brent D. Jordan, "Ovulatory Cycle Effects on Tip Earnings by Lap Dancers: Economic Evidence for Human Estrus?," *Evolution and Human Behavior* 28, no. 6 (November 2007): 375–81, https://doi.org/10.1016/j.evolhumbehav.2007.06.002.

83 **tears:** Shani Gelstein et al., "Human Tears Contain a Chemosignal," *Science* 331, no. 6014 (January 14, 2011): 226–30, https://doi.org/10.1126/science.1198331.

84 **"neither their production, nor their release, [nor their information] content are subject to conscious manipulation":** Katrin T. Lübke et al., "Pregnancy Reduces the Perception of Anxiety," *Scientific Reports* 7, no. 1 (August 23, 2017): 9213, https://doi.org/10.1038/s41598-017-07985-0.

87 **bombykol:** The paper was published in 1961. Bombykol was discovered by Adolf Butenandt, who joined the Nazi Party during World War II and worked on a variety of war science projects. He won the Nobel Prize in 1939 for his work on sex hormones and later headed the prestigious Max Planck Society. Adolf Butenandt, Rüdiger Beckmann, and Erich Hecker, "Über den sexuallockstoff des Seidenspinners, I. Der Biologische Test und die Isolierung des

Reinen Sexuallockstoffes Bombykol," *Biological Chemistry* 324, no. Jahresband (January 1, 1961): 71–83, https://doi.org/10.1515/bchm2.1961.324.1.71.

87   **It works on a vast majority of males, a vast majority of the time:** Biology is inherently chaotic. Nothing is true 100% of the time, and any scientist who says otherwise is probably trying to sell you something. But when it comes to bombykol's ability to attract a male, it is as close to 100% true as biology can come.

87   **male boar:** Wyatt, *Pheromones and Animal Behavior*, 261.

89   **androstenone and androstenol are often found in human sweat:** Wyatt, *Pheromones and Animal Behavior*, 296.

90   **a lot of impotent effort coupled with attempts to sell unproven products:** Alla Katsnelson, "What Will It Take to Find a Human Pheromone?," *ACS Central Science* 2, no. 10 (October 26, 2016): 678–81, https://www.ncbi.nlm.nih .gov/pmc/articles/PMC5084077/.

5: HOT ROCKS

So very many wonderful people spoke to me about sweat bathing. I especially found conversations with Mikkel Aaland, Risto Elomaa, Jari Laukkanen, Tuomo Sarkikoski, Lasse Erikson, Rob Keijzer, and Paolo Dell'Omo to be enlightening.

94   **Dutch spa just outside Amsterdam:** Thermen Soesterberg (website), accessed September 1, 2020, https://www.thermensoesterberg.nl/home.

97   **Aufguss WM:** Aufguss WM (website), accessed September 1, 2020, https:// www.aufguss-wm.com/en/.

99   **the human body is one of the coolest objects in a hot sauna:** Michael Zech et al., "Sauna, Sweat and Science—Quantifying the Proportion of Condensation Water versus Sweat Using a Stable Water Isotope (2H/1H and 18O/16O) Tracer Experiment," *Isotopes in Environmental and Health Studies* 51, no. 3 (July 3, 2015): 439–47, https://doi.org/10.1080/10256016.2015.1057136.

100   **wondered the same thing and conducted studies in 2015:** Zech et al., "Sauna, Sweat and Science."

102   **boost blood levels of epinephrine, growth hormone, and endorphins:** Katriina Kukkonen-Harjula et al., "Haemodynamic and Hormonal Responses to Heat Exposure in a Finnish Sauna Bath," *European Journal of Applied Physiology and Occupational Physiology* 58, no. 5 (March 1, 1989): 543–50, https:// doi.org/10.1007/BF02330710.

103   **"mostly retrospective and poorly controlled":** E. Ernst, "Sauna—A Hobby or for Health?," *Journal of the Royal Society of Medicine* 82, no. 11 (November 1989): 639, https://doi.org/10.1177/014107688908201103.

103   **a more serious study:** Which he published in 1990: E. Ernst et al., "Regular
       Sauna Bathing and the Incidence of Common Colds," *Annals of Medicine* 22,
       no. 4 (January 1990): 225–27, https://doi.org/10.3109/07853899009148930.

103   **"the mean duration and average severity of common colds . . . needed to
       prove this":** Ernst et al., "Regular Sauna Bathing."

104   **excellent for your heart:** Tanjaniina Laukkanen et al., "Association Between
       Sauna Bathing and Fatal Cardiovascular and All-Cause Mortality Events,"
       JAMA Internal Medicine 175, no. 4 (April 1, 2015): 542, https://doi.org/10
       .1001/jamainternmed.2014.8187. The reason I point to this study is because
       it was done on many men over several decades. Many more studies on smaller
       sample sizes have produced conflicting results.

105   **it has been done in hamsters:** Y. Ikeda et al., "Repeated Sauna Therapy
       Increases Arterial Endothelial Nitric Oxide Synthase Expression and Nitric
       Oxide Production in Cardiomyopathic Hamsters," *Circulation Journal* 69,
       no. 6 (June 2005): 722–29, https://doi.org/10.1253/circj.69.722.

106   **sauna multiple times a week:** There have been many sauna science studies with
       conflicting results. I again highlight this one because it had such a large sam-
       ple size: Laukkanen et al., "Association Between Sauna Bathing." The reason
       I point to this study is because it was done on many men over several decades.
       Many more studies on smaller sample sizes have produced conflicting results.

106   **published studies on sweat psychological therapy:** Stephen A. Colmant
       and Rod J. Merta, "Sweat Therapy," *Journal of Experiential Education* 23,
       no. 1 (June 1, 2000): 31–38, https://doi.org/10.1177/105382590002300106;
       and Allen Eason, Stephen Colmant, and Carrie Winterowd, "Sweat Therapy
       Theory, Practice, and Efficacy," *Journal of Experiential Education* 32, no. 2
       (November 1, 2009): 121–36, https://doi.org/10.1177/105382590903200203.

106   **"At first, the heat is soothing . . . deeper state of relaxation":** Eason, Col-
       mant, and Winterowd, "Sweat Therapy Theory, Practice, and Efficacy."

106   **"As the heat becomes more intense . . . faced with adversity":** Eason, Col-
       mant, and Winterowd, "Sweat Therapy Theory, Practice, and Efficacy."

109   **Pakistan:** "Archaeological Ruins at Moenjodaro," UNESCO World Heritage
       Centre, accessed July 23, 2020, https://whc.unesco.org/en/list/138/.

109   **Mexico:** Brigit Katz, "14th-Century Steam Bath Found in Mexico City," *Smith-
       sonian Magazine*, accessed July 23, 2020, https://www.smithsonianmag
       .com/smart-news/14th-century-steam-bath-found-mexico-city-180974049/.

109   **At a recent Global Wellness Summit:** Global Wellness Summit, "8 Well-
       ness Trends for 2017—and Beyond," accessed September 1, 2020, https://
       www.globalwellnesssummit.com/wp-content/uploads/Industry-Research
       /8WellnessTrends_2017.pdf.

110   **a comprehensive history of the Finnish sauna:** Tuomo Särkikoski, *Kiukaan*

kutsu ja löylyn lumo: Suomalaisen saunomisen vuosikymmeniä (Helsinki: Gummerus, 2012).

111 **"worldly fame and reputation are worth less than a rotten lingonberry":** "50 Stunning Olympic Moments No31: Paavo Nurmi Wins 5,000m in 1924 | Simon Burnton," *Guardian*, May 18, 2012, http://www.theguardian.com/sport/blog/2012/may/18/50-stunning-olympic-moments-paavo-nurmi.

111 **archival evidence that Heinrich Himmler:** Särkikoski, *Kiukaan kutsu ja löylyn lumo.*

112 **infrared sauna, a market worth $75 million in 2017:** "What You Need to Know About So-Hot-Right-Now Infrared Spa Therapy," *Bloomberg*, March 24, 2017, https://www.bloomberg.com/news/articles/2017-03-24/what-you-need-to-know-about-so-hot-right-now-infrared-spa-therapy.

119 **Hannu Rautkallio, an earlier Soviet leader, Nikita Khrushchev:** Andrew Osborn, "Nikita in Hot Water for Sauna Frolic," *Guardian*, November 30, 2001, http://www.theguardian.com/world/2001/dec/01/russia.andrewosborn.

6: SWEATPRINTS

This chapter relied on many interviews, particularly with Stephen Bleay, Simona Francese, Jan Halámek, Jayoung Kim, Jill Newton, John Rogers, Juliane Sempionatto, and Joseph Wang.

121 **a case of breaking and entering:** R. Bradshaw, N. Denison, and S. Francese, "Implementation of MALDI MS Profiling and Imaging Methods for the Analysis of Real Crime Scene Fingermarks," *Analyst* 142, no. 9 (2017): 1581–90, https://doi.org/10.1039/C7AN00218A.

121 **chemicals left behind in fingerprints:** S. Francese, R. Bradshaw, and N. Denison, "An Update on MALDI Mass Spectrometry Based Technology for the Analysis of Fingermarks—Stepping into Operational Deployment," *Analyst* 142, no. 14 (2017): 2518–46, https://doi.org/10.1039/C7AN00569E.

121 **traces of cocaine:** Bradshaw, Denison, and Francese, "Implementation of MALDI MS Profiling and Imaging Methods."

121 **cocaethylene:** Peter Jatlow et al., "Alcohol Plus Cocaine: The Whole Is More Than the Sum of Its Parts," *Therapeutic Drug Monitoring* 18, no. 4 (August 1996): 460–64, https://doi.org/10.1097/00007691-199608000-00026.

121 **liver creates a hybrid molecule:** Jatlow et al., "Alcohol Plus Cocaine."

122 **Francis Galton, a cousin of Charles Darwin, popularized the idea:** Gertrud Hauser, "Galton and the Study of Fingerprints," in *Sir Francis Galton, FRS: The Legacy of His Ideas: Proceedings of the Twenty-Eighth Annual Symposium of the Galton Institute, London, 1991*, ed. Milo Keynes. Studies in

Biology, Economy and Society (London: Palgrave Macmillan, 1993), 144–57, https://doi.org/10.1007/978-1-349-12206-6_10.

122 **ninhydrin dye to stain fingerprints:** Svante Oden and Bengt von Hofsten, "Detection of Fingerprints by the Ninhydrin Reaction," *Nature* 173 (March 6, 1954): 449–50.

123 **silver nitrate solutions:** Stephen Bleay, telephone interview by the author, January 2, 2018.

123 **Atomic Energy Authority published a report:** F. Cuthbertson, *The Chemistry of Fingerprints*, AWRE Report No. 013/69 (Aldermaston: UK Atomic Energy Authority, 1969).

123 **perspiration with higher-than-average levels of salt:** People with cystic fibrosis have very salty sweat because the same chloride ion membrane transporters that malfunction in their lungs don't do their job in the eccrine gland to scavenge the salty ion as sweat traverses out of the pore. That leads to higher than normal levels of chloride, which can alert doctors that the person might have cystic fibrosis. Avantika Mishra, Ronda Greaves, and John Massie, "The Relevance of Sweat Testing for the Diagnosis of Cystic Fibrosis in the Genomic Era," *Clinical Biochemist Reviews / Australian Association of Clinical Biochemists* 26, no. 4 (November 2005): 135–53, https://www.ncbi.nlm.nih.gov/pmc/articles/PMC1320177/.

124 **by the mid-2000s, all that began to change:** In 2007, Sergei G. Kazarian and colleagues first proposed using attenuated total reflection Fourier transform infrared (ATR-FT-IR) spectroscopy to image fingerprint chemicals. Camilla Ricci et al., "Chemical Imaging of Latent Fingerprint Residues," *Applied Spectroscopy* 61, no. 5 (May 1, 2007): 514–22, https://doi.org/10.1366/000370207780807849. Since then, MALDI mass spectrometry has become a more widely deployed and developed tool for chemical analysis of fingerprints. See Francese, Bradshaw, and Denison, "An Update on MALDI Mass Spectrometry Based Technology."

124 **hard drugs, such as cocaine, or more benign intoxicants, such as caffeine:** Francese, Bradshaw, and Denison, "An Update on MALDI Mass Spectrometry Based Technology."

124 **gender and age could be identified in the marks:** Francese, Bradshaw, and Denison, "An Update on MALDI Mass Spectrometry Based Technology."

124 **"Sheffield could never pass for a fine town ... outlandish factory-buildings":** François duc de La Rochefoucauld et al., *Innocent Espionage: The La Rochefoucauld Brothers' Tour of England in 1785* (Woodbridge, UK: Boydell & Brewer, 1995).

124 **"What a beautiful place Sheffield would be, if Sheffield were not there!":** Walter White, *A Month in Yorkshire* (London: Chapman & Hall, 1861).

124 **"It could justly claim to be called the ugliest town in the Old World":** "10 of the Funniest Quotes Ever Written About Sheffield," *Sheffield Telegraph*, January 18, 2018, https://www.sheffieldtelegraph.co.uk/read-this/10-funniest -quotes-ever-written-about-sheffield-439032.

124 **"And the stench! If at rare moments you stop smelling sulphur it is because you have begun smelling gas":** George Orwell, *The Complete Works of George Orwell: Novels, Memoirs, Poetry, Essays, Book Reviews & Articles: 1984, Animal Farm, Down and Out in Paris and London, Prophecies of Fascism . . .* (e-artnow, 2019).

125 **cauldron for "synth pop":** Simon Price, "Why Sheffield?," *Guardian*, April 24, 2004, https://www.theguardian.com/music/2004/apr/24/popandrock2.

125 **the monument's pinnacle was struck by lightning in 1990:** Duncan Sayer, *Ethics and Burial Archaeology* (London: Bloomsbury, 2017).

125 **four giant steel buildings shaped like curling stones:** "National Centre for Popular Music—Projects," Nigel Coates, accessed July 8, 2020, https:// nigelcoates.com/projects/project/national_centre_for_popular_music.

126 **stinky emissions of rotting chickens:** "Rotting Chicken Shows Food Emissions Role," BBC News, accessed July 8, 2020, https://www.bbc.com/ news/av/science-environment-34937844/rotting-chicken-shows-food -emissions-role.

126 **time-lapse video of the decomposition:** "Rotting Chicken Shows Food Emissions Role."

129 **vegan or a meat eater:** Jan Havlicek and Pavlina Lenochova, "The Effect of Meat Consumption on Body Odor Attractiveness," *Chemical Senses* 31, no. 8 (October 1, 2006): 747–52, https://doi.org/10.1093/chemse/bjl017.

129 **person's condom brand preferences:** Francese, Bradshaw, and Denison, "An Update on MALDI Mass Spectrometry Based Technology."

130 **men and women release different levels of proteins:** Crystal Huynh et al., "Forensic Identification of Gender from Fingerprints," *Analytical Chemistry* 87, no. 22 (November 17, 2015): 11531–36, https://doi.org/10.1021/acs .analchem.5b03323.

130 **Chemists at the State University of New York:** Huynh et al., "Forensic Identification of Gender."

130 **men and women might be distinguished:** Huynh et al., "Forensic Identification of Gender."

131 **police in the United States and abroad have been collecting biological traces:** Amy Harmon, "Defense Lawyers Fight DNA Samples Gained on Sly," *New York Times*, April 3, 2008, Science, https://www.nytimes.com/2008/04/03/ science/03dna.html.

131 **Advocates of the practice argue:** "When someone leaves his garbage or his biological matter exposed to the public without exerting an effort to retain control

of it, he loses all Fourth Amendment protection for that trash, biological or not," wrote Melanie Baylor, a law professor in the University of Tennessee, Knoxville, here: Melanie D. Wilson, "DNA—Intimate Information or Trash for Public Consumption?," SSRN Scholarly Paper (Rochester, NY: Social Science Research Network, August 31, 2009), https://papers.ssrn.com/abstract=1465043.

132   **"Surreptitious DNA sampling is knocking on the Supreme Court's door. . . . SCOTUS ruling":** Val Van Brocklin, "How Surreptitious DNA Sampling Is Knocking on the Supreme Court's Door," *PoliceOne*, July 29, 2015, https://www.policeone.com/legal/articles/how-surreptitious-dna-sampling-is-knocking-on-the-supreme-courts-door-Aa9RMYXdJbCmYrX2/.

132   **set a precedent for chemical fingerprint analysis:** In addition to surreptitious sampling of DNA, privacy advocates are worried about privacy infringement when forensic scientists test crime-scene DNA for familial connections in genealogical and law enforcement DNA databases. There is less privacy concern about retroactive chemical fingerprint analysis from fingerprint databases because the database entries are a digital image of the fingerprint that is missing chemical information. But any crime-scene objects in storage with fingerprints could be retroactively analyzed.

134   **The next milestone in self-monitoring is chemical:** Amay J. Bandodkar et al., "Wearable Sensors for Biochemical Sweat Analysis," *Annual Review of Analytical Chemistry* 12, no. 1 (June 12, 2019): 1–22, https://doi.org/10.1146/annurev-anchem-061318-114910.

135   **the "holy grail" of chemical surveillance:** Catherine Offord, "Will the Noninvasive Glucose Monitoring Revolution Ever Arrive?," *Scientist*, October 12, 2017, https://www.the-scientist.com/news-analysis/will-the-noninvasive-glucose-monitoring-revolution-ever-arrive-30754.

135   **L'Oréal:** "L'Oréal Unveils Prototype of First-Ever Wearable Microfluidic Sensor to Measure Skin pH Levels," L'Oréal, January 7, 2019, https://mediaroom.loreal.com/en/loreal-unveils-prototype-of-first-ever-wearable-microfluidic-sensor-to-measure-skin-ph-levels/.

136   **Cygnus Incorporated got FDA approval to launch GlucoWatch:** Technically, the FDA gave Cygnus approval for adults 18 and over in 2001 and for children aged 7 to 17 in 2002. "Summary of Safety and Effectiveness Data—GlucoWatch," FDA, August 26, 2002, https://www.accessdata.fda.gov/cdrh_docs/pdf/P990026S008b.pdf.

136   **a small electric current to the skin:** "Summary of Safety and Effectiveness Data—GlucoWatch."

136   **enzymes inside GlucoWatch:** "Summary of Safety and Effectiveness Data—GlucoWatch."

136 **The company did not claim that GlucoWatch was a replacement for finger pricking:** "Summary of Safety and Effectiveness Data—GlucoWatch."

136 **"The *Daily Mail* labeled the device 'a wristwatch to ease diabetes.' . . . finger-prick tests' ":** Offord, "Will the Noninvasive Glucose Monitoring Revolution Ever Arrive?"

136 **"The excitement was tangible":** Offord, "Will the Noninvasive Glucose Monitoring Revolution Ever Arrive?"

136 **the excitement was short-lived:** "The amount of current required to pull glucose out of the skin was enough to cause reddening and burning of the skin (sometimes even blisters), and the accuracy was not good enough to allow it to be used reliably, even as an alarm for low glucose values," wrote diabetes industry consultant John L. Smith in *The Pursuit of Noninvasive Glucose*, 5th ed., 2017, https://www.researchgate.net/publication/317267760_The_Pursuit_of_Noninvasive_Glucose_5th_Edition. Also see: Offord, "Will the Noninvasive Glucose Monitoring Revolution Ever Arrive?"

136 **users got painful rashes:** Offord, "Will the Noninvasive Glucose Monitoring Revolution Ever Arrive?"

136 **not reliable—one study found a false alarm rate of 51%:** The Diabetes Research in Children Network (DirecNet) Study Group, "Accuracy of the GlucoWatch G2 Biographer and the Continuous Glucose Monitoring System During Hypoglycemia. Experience of the Diabetes Research in Children Network (DirecNet)," *Diabetes Care* 27, no. 3 (March 2004): 722–26, https://www.ncbi.nlm.nih.gov/pmc/articles/PMC2365475/.

136 **GlucoWatch production stopped, and its parent company was bought and sold:** David Kliff, "The Return of the 'GlucoWatch,'" *Diabetic Investor* (blog), July 22, 2013, https://diabeticinvestor.com/the-return-of-the-glucowatch-2/.

137 **the design challenges are formidable:** Offord, "Will the Noninvasive Glucose Monitoring Revolution Ever Arrive?"

137 **a drug called pilocarpine:** Donato Vairo et al., "Towards Addressing the Body Electrolyte Environment via Sweat Analysis: Pilocarpine Iontophoresis Supports Assessment of Plasma Potassium Concentration," *Scientific Reports* 7, no. 1 (September 18, 2017): 11801, https://doi.org/10.1038/s41598-017-12211-y.

137 **high levels of pilocarpine have caused seizures in rats:** E. A. Cavalheiro et al., "Long-Term Effects of Pilocarpine in Rats: Structural Damage of the Brain Triggers Kindling and Spontaneous Recurrent Seizures," *Epilepsia* 32, no. 6 (December 1991): 778–82, https://doi.org/10.1111/j.1528-1157.1991.tb05533.x.

137 **arguably the most challenging of all:** John A. Rogers, telephone interview by the author, February 10, 2019.

137 **Other biomarkers in sweat are much easier to track:** Bandodkar et al., "Wearable Sensors for Biochemical Sweat Analysis."

138 **SCRAM Systems launched a sweat-surveillance device:** "About Us," SCRAM Systems, accessed July 9, 2020, https://www.scramsystems.com/our -company/about-us/.

138 **An independent study from researchers at the University of Texas...,** **respectively:** John D. Roache et al., "Using Transdermal Alcohol Monitoring to Detect Low-Level Drinking," *Alcoholism, Clinical and Experimental Research* 39, no. 7 (July 2015): 1120–27, https://doi.org/10.1111/acer.12750.

138 **alcohol levels are measured every 30 minutes:** "What Is the SCRAM CAM Bracelet and How Does It Work?," SCRAM Systems, December 11, 2018, https://www.scramsystems.com/scram-blog/what-is-scram-cam-bracelet -how-does-it-work/.

138 **22,000 people wear their product:** "Counties Augmenting Roadside Checkpoints, Media Campaigns With 24/7 Monitoring to Curb Drunk Drivers," SCRAM Systems, accessed July 9, 2020, https://www.scramsystems.com/ media-room/counties-augmenting-roadside-checkpoints-media-campaigns -with-24-7-monitori/.

138 **760,000 people monitored since the product's launch:** Shauna Rusovick, SCRAM Systems, email correspondence with fact-checker Alla Katsnelson, May 14, 2020.

138 **pop star Lindsay Lohan, actor Tracy Morgan:** John Eligon, "Not Just for the Drunk and Famous: Ankle Bracelets That Monitor Alcohol," *New York Times*, May 30, 2010, https://www.nytimes.com/2010/05/31/nyregion/31ankle.html.

138 **and actress Michelle Rodriguez:** Courtney Rubin, "Michelle Rodriguez Complains About Ankle Bracelet," People.com, February 21, 2007, https://people .com/celebrity/michelle-rodriguez-complains-about-ankle-bracelet/.

139 **"VCR Dog Tag":** Rubin, "Michelle Rodriguez."

139 **a multitude of engineering and chemical hurdles:** Rogers, telephone interview with author.

7: FAKE SWEAT

Many people were kind to provide me with their time: Thanks in particular to Andy Blow, Tamara Hew-Butler, Hein Daanen, Eugene Laverty, Alan McCubbin, the late Michael Pickering, Mayur Rancordas, Sissel Tolaas, and many others.

142 **Chevy low-rider:** Chris Ip, "On the Nose | Engadget," *Engadget* (blog), October 26, 2018, https://www.engadget.com/2018-10-26-on-the-nose-sissel-tolaas -detroit-exhibition.html.

142   **Roman Kaiser has famously hung from a zeppelin:** Roman Kaiser, "Headspace: An Interview with Roman Kaiser," *Future Anterior* 13, no. 2 (2016): 1–9.

144   **a molecule found in cheese:** This cheesy molecule is how Tolaas came upon the idea of making cheese from sweat samples: Given that our skin is populated by trillions of microorganisms, Tolaas wondered if the bacteria making the cheesy scent might be co-opted into making actual cheese. She asked artists and celebrities for samples of sweat, receiving specimens from the likes of Mark Zuckerberg's armpit and Hans Ulrich Obrist's forehead— which she then used to inoculate milk. The best results came from David Beckham (of course), whose sweaty sneaker sweat was turned into a Limburger cheese.

144   **$150 for a bottle:** "Artificial Perspiration 2," Pickering Test Solutions, accessed July 7, 2020, https://www.pickeringtestsolutions.com/artificial -perspiration2/.

144   **Clothing manufacturers:** Michael Pickering, CEO, Pickering Labs, telephone interview by the author, March 12, 2007.

144   **Guitar strings:** Pickering, telephone interview with author.

144   **personal handheld electronics:** Rebecca Smith, email correspondence with author, May 9, 2019.

145   **significant amounts of nickel:** Klara Midander et al., "Nickel Release from Nickel Particles in Artificial Sweat," *Contact Dermatitis* 56, no. 6 (June 15, 2007): 325–30, https://doi.org/10.1111/j.1600-0536.2007.01115.x.

145   **Crime labs also have a steady demand:** Sarah Everts, "Pseudo Sweat," *Newscripts* (blog), March 26, 2007, https://cen.acs.org/articles/85/i13/ Newscripts.html.

146   **"it is safe to say . . . each year":** Smith, email correspondence with author.

146   **"artificial perspiration is our largest selling product category":** Smith, email correspondence with author.

146   **1962 film *Vive le tour*:** Criterion Collection, *Vive le tour—Refueling*, accessed December 11, 2017, https://www.youtube.com/watch?v=2nLxAKwtBb4.

146   **"Here's one of the . . . drink and drink":** Criterion Collection, *Vive le tour—Refueling*.

147   **"Until the 1970s . . . mark of weakness":** Timothy Noakes, *Waterlogged: The Serious Problem of Overhydration in Endurance Sports* (Champaign, IL: Human Kinetics, 2012), xiii.

147   **"to run a complete marathon . . . fitness":** Noakes, *Waterlogged*.

147   **finish their competitive event lighter:** Tamara Hew-Butler, telephone interview by the author, February 8, 2018.

147   **below a hydration level of 15%:** Hew-Butler, telephone interview with author.

148   **five marathon runners died from hyponatremia:** Christie Aschwanden, *Good to Go: What the Athlete in All of Us Can Learn from the Strange Science of Recovery* (New York: W. W. Norton, 2019), 46.

148   **we have a sophisticated water-conservation system:** J. Batcheller, "Disorders of Antidiuretic Hormone Secretion," *AACN Clinical Issues in Critical Care Nursing* 3, no. 2 (1992): 370–78, https://doi.org/10.4037/15597768-1992-2009.

149   **Gatorade, was developed in the 1960s:** "Gatorade Company History," Gatorade, accessed July 7, 2020, http://www.gatorade.com.mx/company/heritage.

149   **water, salt, sugar, citrus flavoring:** Noakes, *Waterlogged*, xvii.

149   **"new population of joggers aspiring to become marathon runners":** Noakes, *Waterlogged*, xvii.

150   **As one dietitian puts it:** Alan McCubbin, telephone interview by the author, December 22, 2017.

150   **caloric heft of most sports . . . 90 minutes:** C. Heneghan et al., "Forty Years of Sports Performance Research and Little Insight Gained," *BMJ* 345 (July 18, 2012): e4797, https://doi.org/10.1136/bmj.e4797.

150   **one beverage company:** Only one manufacturer, GlaxoSmithKline, provided the researchers with a comprehensive bibliography of the trials used to underpin its product claims for Lucozade—a carbohydrate-containing sports drink. "Other manufacturers of leading sports drinks did not supply us with comprehensive bibliographies, and in the absence of systematic reviews we surmise that the methodological issues raised in this article could apply to all other sports drinks," notes the paper (Heneghan et al., "Forty Years of Sports Performance Research").

150   **"If you apply evidence-based methods . . . general public":** Heneghan et al., "Forty Years of Sports Performance Research."

151   **wars fought and expeditions taken for access to the edible crystal:** Mark Kurlansky, *Salt: A World History* (Toronto: Vintage Canada, 2002).

152   **140 millimolar. By the time it reaches the skin, most people are at about 40 millimolar:** Lindsay B. Baker, "Sweating Rate and Sweat Sodium Concentration in Athletes: A Review of Methodology and Intra/Interindividual Variability," *Sports Medicine* 47 (2017): 111–28, https://doi.org/10.1007/s40279-017-0691-5.

151   **double or triple the amount:** Baker, "Sweating Rate and Sweat Sodium Concentration."

152   **What has been published suggests athletes should replace lost electrolytes gradually:** Telephone interview with Alan McCubbin, Zoom follow-up with fact-checker, July 31, 2020.

152   **naked athletes in large plastic bags:** S. M. Shirreffs and R. J. Maughan,

"Whole Body Sweat Collection in Humans: An Improved Method with Preliminary Data on Electrolyte Content," *Journal of Applied Physiology* 82, no. 1 (January 1, 1997): 336–41, https://doi.org/10.1152/jappl.1997.82.1.336.

153 **a drug called pilocarpine . . . activate sweating:** Donato Vairo et al., "Towards Addressing the Body Electrolyte Environment via Sweat Analysis: Pilocarpine Iontophoresis Supports Assessment of Plasma Potassium Concentration," *Scientific Reports* 7, no. 1 (2017): 11801, https://doi.org/10.1038/s41598-017-12211-y.

154 **unpalatable:** Hew-Butler, telephone interview with author.

154 **"there's no reason . . . hydrating brew":** Aschwanden, *Good to Go*, 35–36.

155 **"The research suggests . . . drink it":** Aschwanden, *Good to Go*.

8: A ROSE BY ANY OTHER NAME

This chapter benefited from many conversations and interviews, in particular those with Isabelle Chazot, Jean Kerléo, Lauryn Mannigel, Eugénie Briot, Cecilia Bembibre, Donna Bilak, and Philippe Walter.

159 **we've used perfume to modify the scent:** Constance Classen, David Howes, and Anthony Synnott, *Aroma: The Cultural History of Smell* (London: Routledge, 1994); Alain Corbin, *The Foul and the Fragrant* (New York: Berg, 1986).

159 **bathed our bodies:** Katherine Ashenburg, *The Dirt on Clean: An Unsanitized History* (New York: North Point Press, 2007).

159 **Humans have also worn perfume as protection:** We put pleasant odors in "smell boxes," as described by Corbin, *The Foul and the Fragrant*.

159 **Egyptian limestone relief from around 600 BCE:** *Lintel from the Tomb of Païrkep with Bas-Relief Sculpture: Making Lily Perfume*, règne de Psammétique II ?(—589 avant J.-C.), 26e dynastie 595, calcaire, H. 0.29 m; W. 1.1 m; D. 0.08 m, règne de Psammétique II ?(—589 avant J.-C.), 26e dynastie 595, Louvre, https://www.louvre.fr/en/oeuvre-notices/lintel-tomb-pairkep-bas-relief-sculpture-making-lily-perfume.

160 *kyphi,* **whose 16 ingredients:** Classen, Howes, and Synnott, *Aroma*.

160 **"When we think of . . . for our ancestors":** Classen, Howes, and Synnott, *Aroma*.

160 **4,000-year-old perfume factory:** Malcolm Moore, "Eau de BC: The Oldest Perfume in the World," *Telegraph*, March 21, 2007, https://www.telegraph.co.uk/news/worldnews/1546277/Eau-de-BC-the-oldest-perfume-in-the-world.html.

160 **[He] steeps his feet . . . ground thyme:** Eugene Rimmel, *The Book of Per-*

fumes, 5th ed. (London: Chapman & Hall, 1867), https://archive.org/details/
bookofperfumes00rimm/page/84/mode/2up/search/egyptian+unguents.

161 **fragrant dishes to be found in the Mediterranean region:** Classen, Howes,
and Synnott, *Aroma*.

161 **retired perfumer Jean Kerléo:** Jean Kerléo, in-person interview at Osmothèque by the author, January 16, 2018.

163 **Pliny the Elder's recipe had listed 27 ingredients:** Jean Kerléo, "Un Parfum
Romain: Le Parfum Royal." Unpublished manuscript.

164 **a tenth-century Arab discovery:** Peter Burne, *The Teetotaler's Companion;
Or, A Plea for Temperance* (London: Arthur Hall, 1847).

164 **from *the* Cologne, Germany:** The original Eau de Cologne was invented by
Giovanni Maria Farina, an Italian immigrant perfumer in the city, in 1709. It
quickly became so popular that the city's name became a moniker for scented
extracts or essential oils mixed with alcohol and water. "Original Eau de
Cologne Celebrates 300 Years | DW | 13.07.2009," *Deutsche Welle* (blog), July
13, 2009, https://www.dw.com/en/original-eau-de-cologne-celebrates-300
-years/a-4475632.

166 **the Industrial Revolution and scientific discoveries:** Eugénie Briot,
"From Industry to Luxury: French Perfume in the Nineteenth Century,"
Business History Review 85, no. 2 (2011): 273–94, https://doi.org/10.1017/
S0007680511000389.

166 **flowers were exposed to grease or oil:** "Perfume," in *Encyclopedia Britan-
nica*, accessed July 10, 2020, https://www.britannica.com/art/perfume.

166 **rational saturator, some 800 kilograms:** Briot, "From Industry to Luxury."

166 **synthetic chemists figured out ways to produce odors:** Briot, "From Industry
to Luxury."

167 **Spanish explorer Hernán Cortés purportedly witnessed the Aztec emperor
Montezuma:** Henry B. Heath, *Source Book of Flavors*. AVI Sourcebook and
Handbook Series (New York: Van Nostrand Reinhold, 1981).

167 **vanillin made an appearance in Jicky:** Patricia de Nicolaï, "A Smelling Trip
into the Past: The Influence of Synthetic Materials on the History of Per-
fumery," *Chemistry & Biodiversity* 5, no. 6 (June 2008): 1137–46, https://doi
.org/10.1002/cbdv.200890090.

167 **"Lest anyone think that unisex perfumes . . . per hour":** Luca Turin and
Tania Sanchez, *Perfumes: The A–Z Guide* (New York: Penguin Books, 2009).

168 **heliotrope was one of the most desired:** Briot, "From Industry to Luxury."

168 **the upper class abandoned the scent:** Briot, "From Industry to Luxury."

168 **similar aromas were put in beautiful bottles:** Briot, "From Industry to
Luxury."

9: ARMING THE ARMPIT

I am particularly grateful for interviews with Chris Callewaert, Cari Casteel, Ariane Lenzner, and Juliann Sivulka.

169 **trying unsuccessfully to promote an antiperspirant:** "Odorono Company 1925–1936. Account Histories," Box 33, JWT Corporate Archives records, Hartman Center for Marketing Advertising and History, David M. Rubenstein Rare Book & Manuscript Library, Duke University, accessed May 8, 2012.

169 **Borrowing $150 from her grandfather:** "Odorono Company 1925–1936. Account Histories."

169 **"This was still very much a Victorian society . . . bodily functions, in public":** Juliann Sivulka, telephone interview by the author, April 28, 2012.

169 **most people's solution to body odor was to wash regularly with soap and water:** Katherine Ashenburg, *The Dirt on Clean: An Unsanitized History*, (New York: North Point Press, 2007).

170 **baking soda:** Abby Slocomb and Jennie Day, Deodorizing Perspiration Powder, US Patent Office 279195 (New Orleans, Louisiana, filed December 26, 1882, and issued June 12, 1883).

170 **cayenne:** Sam Clayton, Improved Medical Compound, US Patent Office 52032 (South Amboy, New Jersey, n.d.).

170 **formaldehyde:** Henry Blackmore, Formaldehyde Product and Process of Making Same, US Patent Office 795757 (Mount Vernon, New York, filed September 4, 1904, and issued July 25, 1905); and Armand Gardos, Treated Stocking, US Patent Office 1219451 (Cleveland, Ohio, filed October 4, 1915, and issued March 20, 1917).

170 **One of the first US patents:** Henry D. Bird, Improved Compound for Cleansing the Human Body from Offensive Odors, US Patent Office 64189 (Petersburg, Virginia, issued April 20, 1867).

170 **"I am aware . . . . odor from the human body":** Bird, "Improved Compound."

170 **baker's yeast:** George T. Southgate, Deodorant Composition, US Patent Office 1729752 (Forest Hills, New York, filed February 23, 1926, and issued October 29, 1929).

170 **"the greater vitality of the yeast fermentation than that of putrefaction":** Southgate, "Deodorant Composition."

170 **The first trademarked deodorant was launched in 1888:** Specifically, the product's name Mum was used starting in 1888, according to the 1905 trademark documents here: "MUM Trademark—Registration Number 0072837—

Serial Number 71038770 : Justia Trademarks," accessed July 12, 2020, http://
tmsearch.uspto.gov/bin/showfield?f=doc&state=4806:al8z5u.10.1.

170 **zinc oxide to destroy armpit bacteria:** Karl Laden, *Antiperspirants and
Deodorants,* 2nd ed. (Boca Raton, FL: CRC Press, 1999).

170 **The first trademarked antiperspirant, Everdry:** Laden, *Antiperspirants
and Deodorants.*

171 **Coolene:** Coolene was a product of Coolene Company, a manufacturer of toiletries that operated from at least 1900 to 1917 and obtained a patent for its
product bottle in 1904: Harry G. Lord, Bottle, United States 777477A (filed
August 30, 1904, and issued December 13, 1904), https://patents.google.com
/patent/US777477/en.

171 **"medicated toilet delight":** *Advertisement for Coolene,* early twentieth century, Sarah Everts's Collection of Vintage Advertisements.

171 **"imperfections in your toilet . . . with perspiration excreta":** *Advertisement
for Coolene.*

171 **"The [exhibition] demonstrator could not sell any Odorono . . . cover
expenses":** "Odorono Company 1925–1936. Account Histories."

171 **door-to-door Bible salesman:** "Sidney Ralph Bernstein Company History
Files, 1873–1964," n.d., Box 5, JWT Corporate Archives records, Hartman
Center for Marketing Advertising and History, David M. Rubenstein Rare
Book & Manuscript Library, Duke University, accessed May 9, 2012.

171 **one of the most famous advertising copywriters:** James Webb Young was
inducted into the Advertising Hall of Fame in 1974: "Members: James Webb
Young, 1886–1973, Inducted 1974," Advertising Hall of Fame, accessed July 12,
2020, http://advertisinghall.org/members/member_bio.php?memid=826.

172 **Odorono stopped sweat for up to 3 days:** "Odorono Company 1925–1936.
Account Histories."

172 **strong acid to remain effective:** Laden, *Antiperspirants and Deodorants.*

172 **"violent irritant . . . 'perspiration preventive' ":** Council on Pharmacy and
Chemistry and the Association Laboratory, "Propaganda for Reform: ODOR-
O-NO," *Journal of the American Medical Association* LXII, no. 1 (January 3,
1914): 54, https://doi.org/10.1001/jama.1914.02560260062031.

172 **"guaranteed by the manufacturer to be absolutely harmless":** Council on
Pharmacy and Chemistry and the Association Laboratory, "Propaganda for
Reform."

172 **Odorono solution was red:** Council on Pharmacy and Chemistry and the
Association Laboratory, "Propaganda for Reform."

172 **ruined many a fancy outfit, including one woman's wedding dress:** "Odorono Company 1925–1936. Account Histories."

172   **Odorono advised its customers:** "Odorono Company 1925–1936. Account
        Histories."

172   **"excessive perspiration":** Juliann Sivulka, "Odor, Oh No! Advertising
        Deodorant and the New Science of Psychology, 1910 to 1925," in *Proceedings
        of the 13th Conference on Historical Analysis & Research in Marketing*, ed.
        Blaine J. Branchik (CHARM Association, 2007), 212–20.

172   **But by 1919, Odorono's market upswing had flattened:** Sivulka, "Odor, Oh
        No!" Specifically, the flattening was noticed by 1918, and a survey was launched
        to find the cause of the flattening, and by 1919 the pressure mounted on Young.

173   **"every woman knew of Odorono and about one-third used the product":**
        "Odorono Company 1925–1936. Account Histories."

173   **"Within the Curve of a Woman's arm. A frank discussion of a subject too
        often avoided":** You can see the advertisement in many places. I suggest:
        Sivulka, "Odor, Oh No!"

173   **"A woman's arm! . . . it isn't, always":** Sivulka, "Odor, Oh No!"

173   **Young's memoir:** "Odorono Company 1925–1936. Account Histories."

173   **Odorono sales rose 112% within a year, to $417,000:** "Odorono Company
        1925–1936. Account Histories."

173   **she sold the company to Northam Warren:** Edna Patricia Murphey's life
        as an entrepreneur continued, albeit under her new name, Patricia Winter,
        after she married the artist Ezra Winter in 1932. "Her career as a cosmet-
        ics mogul behind her, she retired from active duty, only to reinvent herself
        once again—as a farmer, who would go on to build an herbal empire that she
        would sell in the 1950s to McCormick Spices," as noted here: Jessica Hel-
        fand, "Ezra Winter Project: Chapter Four," *Design Observer*, November
        30, 2016, http://designobserver.com/feature/ezra-winter-project-chapter
        -four/33818.

174   **"Beautiful but dumb. She has never learned the first rule of lasting charm":**
        Helfand, "Ezra Winter Project: Chapter Four."

174   **"Why will so many married women consider themselves so safe?":** "Odor-
        ono Company 1925–1936. Account Histories."

174   **"Is it that they are blind . . . safely married?":** "Odorono Company 1925–
        1936. Account Histories."

174   **"began adding snarky comments . . . buy, buy two":** Cari Casteel, telephone
        interview by the author, July 3, 2012.

175   **"I consider a body deodorant for masculine use to be sissified":** "Odorono
        Company 1925–1936. Account Histories."

175   **"I feel there is a market for deodorants among men . . . leading men's maga-
        zine?":** "Odorono Company 1925–1936. Account Histories."

175   **Top-Flite, and sold it for 75 cents:** "Top-Flite Advertisement in Life Magazine," *Life*, 1935, p. 43, https://books.google.ca/books/content?id=et9GAAAAMAAJ &pg=RA11-PA43&img=1&zoom=3&hl=en&sig=ACfU3U0WnB4IBYKPjS1xLQz B71wiHiDF3w&ci=45%2C19%2C897%2C1243&edge=0.

175   **"The Depression shifted . . . advertisement said":** Casteel, telephone interview with author.

175   **"because the company owner Alfred McKelvy said he 'couldn't think of anything more manly than whiskey' ":** Casteel, telephone interview with author.

175   **"a special vocabulary":** Chris Welles, "Big Boom in Men's Beauty Aids: Not by Soap Alone," *Life*, August 13, 1965, https://books.google.ca/books?id=MVME AAAAMBAJ&pg=PA39&dq=deodorant+special+vocabulary+1965&hl=en&sa =X&ved=2ahUKEwiIj4f298rqAhVCmXIEHQTvCZ8Q6AEwAH0ECAUQAg#v =onepage&q&f=false.

175   **"using adjectives like tantalizing, crisp, suave, vigorous, robust, virile, and manly":** Welles, "Big Boom in Men's Beauty Aids."

176   **"Big Boom in Men's Beauty Aids":** Welles, "Big Boom in Men's Beauty Aids."

176   **20% of the US cosmetics market:** Welles, "Big Boom in Men's Beauty Aids."

176   **$75 billion industry:** M. Shahbandeh, "Size of the Global Antiperspirant and Deodorant Market 2012–2025," Statista, accessed July 14, 2020, https:// www.statista.com/statistics/254668/size-of-the-global-antiperspirant-and -deodorant-market/.

176   **acid was essential for stabilizing aluminum chloride:** Laden, *Antiperspirants and Deodorants*.

176   **a third molecule:** Jules B. Montenier, Astringent Preparation, US Patent Office 2230083 (Chicago, Illinois, filed December 18, 1939, and issued January 28, 1941).

177   **Montenier patented a plastic squeeze bottle dispenser:** Jules B. Montenier, Unitary Container and Atomizer for Liquids, US Patent Office 2642313 (Chicago, Illinois, filed October 27, 1947, and issued June 16, 1953).

177   **"misted":** Vintage Fanatic, *Stopette Spray Deodorant Commercial 1952*, accessed May 11, 2019, https://www.youtube.com/watch?v=w1Q1rVV5wsk.

177   **if you believe the vintage television advertisements:** Vintage Fanatic, *Stopette Spray Deodorant*.

177   **aluminum chlorohydrate:** Carl N. Andersen, Aluminum Chlorohydrate Astringent, US Patent Office 2492085 (New York, filed May 6, 1947, and issued December 20, 1949).

177   **one of this market's most important technological breakthroughs:** Laden, *Antiperspirants and Deodorants*.

177   **keep pits dry for days:** This is why many of the prescription-strength antiper-

spirants on the market today rely on more acidic solvents and good old aluminum chloride. "Odorono Company 1925–1936. Account Histories."

178   **Helen Barnett decided to fix:** "Death Notice: Helen Barnett," *New York Times*, April 17, 2008, https://archive.nytimes.com/query.nytimes.com/gst/ fullpage-9F05E0D9103AF934A25757C0A96E9C8B63.html.

178   **the first roll-on deodorant:** "Death Notice: Helen Barnett," *New York Times*. It's worth noting that the patent was actually issued to the Bristol-Myers Company's legal assignor, Ralph Henry Thomas, Dispenser, US Patent Office 2749566 (Rahway, New Jersey, filed September 4, 1952, and issued June 12, 1956).

178   **Ban, was a hit:** Anthony Ramirez, "All About/Deodorants; The Success of Sweet Smell," *New York Times*, August 12, 1990, Business Day, https://www .nytimes.com/1990/08/12/business/all-about-deodorants-the-success-of -sweet-smell.html.

178   **technology that exploded out of the starting gate and then crashed in the 1970s:** Laden, *Antiperspirants and Deodorants*, 9–12.

178   **patented in 1941 for dispensing insecticidal bug spray:** Lyle D. Goodhue and William N. Sullivan, Dispensing Apparatus, US Patent Office 2331117A (filed October 3, 1941, and issued October 5, 1943), https://patents.google.com/ patent/US2331117/en.

178   **"the aerosol format for underarm applications. . . products":** Laden, *Antiperspirants and Deodorants*.

178   **pressurized cans could explode:** "The Dangers That Come in Spray Cans," *Changing Times: Kiplinger's Personal Finance*, August 1975, https://www .google.ca/search?tbm=bks&hl=en&q=Changing+Times%2C+%E2%80%9CTh e+company+discovered%2C+and+told+FDA%2C+that+monkeys+exposed+to+ the+sprays+developed+inflamed+lungs.%E2%80%9D.

179   **In 1973, soon after Gillette . . . monkeys exposed to the sprays developed inflamed lungs":** "The Dangers That Come in Spray Cans"; and Laden, *Antiperspirants and Deodorants*, 9–12.

179   **Sesame seeds, spinach, and potatoes . . . thyme, oregano, and chili powder:** Norwegian Scientific Committee for Food Safety, *Risk Assessment of the Exposure to Aluminium Through Food and the Use of Cosmetic Products in the Norwegian Population*. VKM Report 2013 (Oslo: Norwegian Food Safety Authority, 2013), 24, 61.

180   **aluminum phosphate and sodium aluminum sulfate:** Maged Younes et al., "Re-Evaluation of Aluminium Sulphates (E 520–523) and Sodium Aluminium Phosphate (E 541) as Food Additives," *EFSA Journal* 16, no. 7 (2018): e05372, https://doi.org/10.2903/j.efsa.2018.5372.

180   **Most of the aluminum we ingest . . . expunge it in pee:** When you eat food containing trace aluminum, much of the metal passes right through and gets

pooped out. But when trace aluminum is absorbed into the bloodstream, the primary way out is in pee via the kidneys. "Public Health Statement: Aluminum," Agency for Toxic Substances and Disease Registry, September 2008, https://www.atsdr.cdc.gov/ToxProfiles/tp22-c1-b.pdf; Rianne de Ligt et al., "Assessment of Dermal Absorption of Aluminum from a Representative Antiperspirant Formulation Using a 26Al Microtracer Approach," *Clinical and Translational Science* 11, no. 6 (November 2018): 573–81, https://doi.org/10.1111/cts.12579; Calvin C. Willhite et al., "Systematic Review of Potential Health Risks Posed by Pharmaceutical, Occupational and Consumer Exposures to Metallic and Nanoscale Aluminum, Aluminum Oxides, Aluminum Hydroxide and Its Soluble Salts," *Critical Reviews in Toxicology* 44, suppl. 4 (October 2014): 1–80, https://doi.org/10.3109/10408444.2014.934439.

180   **30 to 50 milligrams of aluminum:** Norwegian Scientific Committee for Food Safety, *Risk Assessment of the Exposure to Aluminium*, 17.

180   **2 milligrams per kilogram of body weight per week:** Norwegian Scientific Committee for Food Safety, *Risk Assessment of the Exposure to Aluminium*, 11.

180   **kidney dialysis, some patients were accidentally poisoned:** Allen C. Alfrey, Gary R. LeGendre, and William D. Kaehny, "The Dialysis Encephalopathy Syndrome," *New England Journal of Medicine* 294, no. 4 (January 22, 1976): 184–88, https://doi.org/10.1056/NEJM197601222940402; and "Dialysis Dementia," *British Medical Journal* 2, no. 6046 (November 20, 1976): 1213–14, https://doi.org/10.1136/bmj.2.6046.1213.

180   **Many studies have disproven this theory:** Willhite et al., "Systematic Review of Potential Health Risks."

180   **explanations on their websites:** Alzheimer's Association (USA), "Myths About Alzheimer's Disease," Alzheimer's Disease and Dementia, accessed July 13, 2020, https://alz.org/alzheimers-dementia/what-is-alzheimers/myths.

181   **According to the best available evidence, assessed in 2020 by European risk authorities, using aluminum antiperspirants does not pose a risk to your health:** Scientific Committee on Consumer Safety, *Opinion on the Safety of Aluminium in Cosmetic Products* (Luxembourg: European Commission, March 4, 2020), https://ec.europa.eu/health/sites/health/files/scientific_committees/consumer_safety/docs/sccs_o_235.pdf.

181   **three at the point of publication:** The first in 2001: R. Flarend et al., "A Preliminary Study of the Dermal Absorption of Aluminium from Antiperspirants Using Aluminium-26," *Food and Chemical Toxicology* 39, no. 2 (February 2001): 163–68, https://doi.org/10.1016/S0278-6915(00)00118-6. The second at the request of French health authorities was conducted by Alain Pineau, Olivier Guillard, and colleagues and reported here: Alain Pineau et al., "In

Vitro Study of Percutaneous Absorption of Aluminum from Antiperspirants Through Human Skin in the Franz™ Diffusion Cell," *Journal of Inorganic Biochemistry* 110 (May 2012): 21–26, https://doi.org/10.1016/j.jinorgbio.2012 .02.013. And finally, the Scientific Committee on Consumer Safety commissioned research of its own, which is described in: Scientific Committee on Consumer Safety, *Opinion on the Safety of Aluminium in Cosmetic Products*(Luxembourg: European Commission, March 4, 2020) (with the first portion of that research published in de Ligt et al., "Assessment of Dermal Absorption of Aluminum").

I should point out that there was a study, in 1958, of aluminum penetration through excised skin, but I have not been able to get access to it, and I see no evidence that the researchers attempted to measure subsequent *body burden* of the metal. I'm also skeptical that the analytical equipment of that era was sensitive enough to assess body burden, even if that had been a goal. See I. H. Blank, J. L. Jones, and E. Gould, "A Study of the Penetration of Aluminum Salts into Excised Skin," *Proceedings of the Scientific Section of the Toilet Goods Association* 29 (1958): 32–35.

181 **For a long time, public-health-risk analysts . . . whether this was a reasonable assumption:** Ariane Lenzner, head of aluminum analysis at the German Federal Institute for Risk Assessment, in-person interview by the author, December 4, 2018.

181 **In 2001, the first such experiment took place:** R. Flarend et al., "A Preliminary Study."

182 **"a one-time use of aluminium chlorohydrate . . . body burden of aluminium":** R. Flarend et al., "A Preliminary Study."

182 **in 2007, France's federal agency responsible for the safety of health products:** The study resulted in this 2011 report: French Health Products Safety Agency, *Risk Assessment Related to the Use of Aluminum in Cosmetic Products*, October 2011, https://www.ansm.sante.fr/var/ansm_site/storage/original/application/bfd7283f781cd5ce7d59c151c714ba32.pdf.

182 **the French researchers used a proxy instead:** Pineau et al., "In Vitro Study."

182 **wasn't particularly enlightening:** Pineau et al., "In Vitro Study." It is also worth noting that four of the original nine authors were removed from the list of authors in a subsequent correction—a highly unusual occurrence. Alain Pineau et al., "Corrigendum to 'In Vitro Study of Percutaneous Absorption of Aluminum from Antiperspirants Through Human Skin in the Franz™ Diffusion Cell' [J Inorg Biochem 110 (2012) 21–26]," *Journal of Inorganic Biochemistry* 116 (November 2012): 228, https://doi.org/10.1016/j.jinorgbio.2012.05.014.

The European Commission's Scientific Committee on Consumer Safety called the study "limited" and noted "many other shortcomings." Scientific

Committee on Consumer Safety, *2014 Opinion on the Safety of Aluminium in Cosmetic Products* (Luxembourg: European Commission, March 2014).

183 **The study, published in 2012:** Pineau et al., "In Vitro Study."

183 **French medical regulatory agency to sound an alarm:** French Health Products Safety Agency, *Risk Assessment.*

183 **Norwegian:** Norwegian Scientific Committee for Food Safety, *Risk Assessment of the Exposure to Aluminium*, 17.

183 **German regulators:** *Aluminium-Containing Antiperspirants Contribute to Aluminium Intake* (Berlin: Federal Institute for Risk Assessment, 2014).

183 **the European Union's Scientific Committee on Consumer Safety:** The committee is tasked with providing the EU with "scientific advice it needs when preparing policy and proposals relating to consumer safety, public health and the environment."

183 **too many scientific shortcomings:** Scientific Committee on Consumer Safety, *2014 Opinion on the Safety of Aluminium in Cosmetic Products* (Luxembourg: European Commission, March 2014).

183 **Finally in 2020, the SCCS adopted a final assessment from experiments performed in 18 humans:** Scientific Committee on Consumer Safety, *Opinion on the Safety of Aluminium in Cosmetic Products* (Luxembourg: European Commission, March 4, 2020).

183 **"systemic exposure to aluminium via daily applications of cosmetic products does not add significantly to the systemic body burden of aluminium from other sources":** Scientific Committee on Consumer Safety, *Opinion on the Safety of Aluminium in Cosmetic Products* (Luxembourg: European Commission, March 4, 2020).

184 **Chris Callewaert:** Chris Callewaert, in-person interview by the author, August 13, 2018.

184 **"there are more bacteria in your armpit than humans on this planet, so you should never feel alone":** *Fighting Against Smelly Armpits: Chris Callewaert at TEDxGhent*, 2013, video, https://www.youtube.com/watch?v=9RIFyqLXdVw.

185 *Corynebacterium* **living in your armpits:** A. Gordon James et al., "Microbiological and Biochemical Origins of Human Axillary Odour," *FEMS Microbiology Ecology* 83, no. 3 (2013): 527–40, https://doi.org/10.1111/1574-6941 .12054; and Chris Callewaert et al., "Characterization of *Staphylococcus* and *Corynebacterium* Clusters in the Human Axillary Region," *PLOS ONE* 8, no. 8 (August 12, 2013): e70538, https://doi.org/10.1371/journal.pone.0070538.

186 **hand microbiome swallows up the newcomers:** RadioLab, "The Handshake Experiment | Only Human," WNYC Studios, accessed July 14, 2020, https://www.wnycstudios.org/podcasts/onlyhuman/episodes/handshake -experiment.

187 **identical twins:** Chris Callewaert, Jo Lambert, and Tom Van de Wiele, "Towards a Bacterial Treatment for Armpit Malodour," *Experimental Dermatology* 26, no. 5 (2017): 388–91, https://doi.org/10.1111/exd.13259.

188 **as rare bacteria such as *Anaerococcus* overcompensate for their low numbers by producing extremely potent odors:** Callewaert, in-person interview with author.

189 **"It had taken me a month to coax a new colony of bacteria onto my body. It took me three showers to extirpate it. Billions of bacteria, and they had disappeared as invisibly as they arrived":** Julia Scott, "My No-Soap, No-Shampoo, Bacteria-Rich Hygiene Experiment," *New York Times Magazine*, May 25, 2014, https://www.nytimes.com/2014/05/25/magazine/my-no-soap -no-shampoo-bacteria-rich-hygiene-experiment.html.

190 **Cosmetic company scientists are eyeing the bacterial enzymes used by these microorganisms to turn mostly odorless sweat into dank aroma:** Laden, *Antiperspirants and Deodorants.*

190 **Capture the stinky odors in tiny little molecular cages:** Laden, *Antiperspirants and Deodorants.*

190 **encapsulated in a 2012 art project:** Lucy McRae, *Swallowable Parfum*, 2011, video, https://vimeo.com/27005710.

190 **TED Talk by Lucy McRae:** Lucy McRae, "How Can Technology Transform the Human Body?," 2012, https://www.ted.com/talks/lucy_mcrae_how_can_ technology_transform_the_human_body/transcript.

191 **"a cosmetic pill . . . an atomizer":** McRae, "How Can Technology Transform."

10: EXTREME SWEAT

Much of this chapter was made possible thanks to conversations and correspondence with people who have hyperhidrosis, in particular Mikkel Bjerregaard, Cath Ford, Maria Thomas, Brandon Woodard, Alex Blynn, and the thousand-plus members of the ETS Facebook support group. I also thank Christoph Schick and John Langenfeld for discussions about the history and practice of ETS.

193 **15 million Americans have hyperhidrosis:** Shiri Nawrocki and Jisun Cha, "The Etiology, Diagnosis, and Management of Hyperhidrosis: A Comprehensive Review: Etiology and Clinical Work-Up," *Journal of the American Academy of Dermatology* 81, no. 3 (2019): 657–66, https://doi.org/10.1016/j.jaad .2018.12.071.

193 **"triggers, including crowded areas, emotional provocations, spicy foods, and alcohol":** Shiri Nawrocki and Jisun Cha, "The Etiology, Diagnosis, and Management of Hyperhidrosis: Therapeutic Options," *Journal of the Amer-*

ican Academy of Dermatology 81, no. 3 (2019): 669–80, https://doi.org/10
.1016/j.jaad.2018.11.066.

194   **63% felt unhappy or depressed about their sweating, and 74% felt emotionally damaged from the condition:** Henning Hamm et al., "Primary Focal
Hyperhidrosis: Disease Characteristics and Functional Impairment," *Dermatology* 212, no. 4 (2006): 343–53, https://doi.org/10.1159/000092285.

194   **"I found Uriah . . . a snail":** Charles Dickens, *David Copperfield*, unabridged
ed. CreateSpace Independent Publishing Platform.

194   **a genetic component:** Nawrocki and Cha, "The Etiology, Diagnosis, and Management of Hyperhidrosis: Etiology and Clinical Work-Up."

194   **there is nothing unusual about sweat gland size, shape, or quantity:** Nawrocki and Cha, "The Etiology, Diagnosis, and Management of Hyperhidrosis:
Etiology and Clinical Work-Up."

194   **problematic signaling of the autonomic nervous system:** Nawrocki and Cha,
"The Etiology, Diagnosis, and Management of Hyperhidrosis: Etiology and
Clinical Work-Up."

195   **aberrant central control of emotions:** Nawrocki and Cha, "The Etiology, Diagnosis, and Management of Hyperhidrosis: Etiology and Clinical Work-Up."

195   **a tenth and a fifth of a teaspoon of sweat per minute:** Nawrocki and Cha,
"The Etiology, Diagnosis, and Management of Hyperhidrosis: Etiology and
Clinical Work-Up."

195   **close to three tablespoons per minute:** Nawrocki and Cha, "The Etiology, Diagnosis, and Management of Hyperhidrosis: Etiology and Clinical
Work-Up."

196   **turn of the twentieth century:** Kevin Y. C. Lee and Nick J. Levell, "Turning
the Tide: A History and Review of Hyperhidrosis Treatment," *JRSM Open* 5,
no. 1 (2014): 2042533313505511, https://doi.org/10.1177/2042533313505511.

197   **treat epilepsy, goiters, angina, and glaucoma:** M. Hashmonai and D. Kopelman, "History of Sympathetic Surgery," *Clinical Autonomic Research* 13
(2003): i6–i9 https://doi.org/10.1007/s10286-003-1103-5.

197   **a Swiss medical journal in 1920:** Anastas Kotzareff, "Résection partielle de
tronc sympathique cervical droit pour hyperhidrose unilatérale," *Revue med-
icale de la Suisse Romande* 40 (1920): 111–13.

197   **American doctor named Alfred Adson:** Alfred W. Adson, Winchell McK.
Craig, and George E. Brown, "Essential Hyperhidrosis Cured by Sympathetic
Ganglionectomy and Trunk Resection," *Archives of Surgery* 31, no. 5 (1935):
794–806, https://doi.org/10.1001/archsurg.1935.01180170119008.

197   **"Patients find it impossible . . . embarrassing situations":** Adson, Craig,
and Brown, "Essential Hyperhidrosis Cured."

197 **wasn't widely adopted until minimally invasive procedures were designed in the 1990s:** Hashmonai and Kopelman, "History of Sympathetic Surgery."

198 **"It takes about 10 minutes a side":** *Hyperhidrosis HealthTalk Featuring Dr. John Langenfeld*, 2018, video, https://www.youtube.com/watch?v=NE ReytT2kOg.

198 **"I'd like to do . . . compensatory sweating which could happen":** *Hyperhidrosis HealthTalk*. Langenfeld was surprised to hear of the skepticism from hyperhidrosis patient advocacy groups about ETS surgery. "If I felt overwhelmed that there were too many people coming to me with a severe problem [with compensatory sweating], I would have stopped doing [the operation]," he said in a Zoom interview on October 8, 2020, adding that maybe the long-term outcomes of ETS surgery should be further investigated.

198 **Most everybody who undergoes ETS surgery gets some compensatory sweating:** Antti Malmivaara et al., "Effectiveness and Safety of Endoscopic Thoracic Sympathectomy for Excessive Sweating and Facial Blushing: A Systematic Review," *International Journal of Technology Assessment in Health Care* 23, no. 1 (2007): 54–62, https://doi.org/10.1017/S0266462307051574.

198 **90% of hyperhidrosis patients who received ETS surgery between 1966 and 2004 got some compensatory sweating:** Malmivaara et al., "Effectiveness and Safety of Endoscopic Thoracic Sympathectomy."

199 **4% of ETS surgical patients regretted . . . 11% of patients had regrets:** José Ribas Milanez de Campos et al., "Quality of Life, Before and After Thoracic Sympathectomy: Report on 378 Operated Patients," *Annals of Thoracic Surgery* 76, no. 3 (September 2003): 886–91, https://doi.org/10.1016/S0003 -4975(03)00895-6.

199 **Facebook support group:** "(1) ETS (Endoscopic Thoracic Sympathectomy): Side-Effects, Awareness, & Support | Facebook," accessed January 17, 2021, https://www.facebook.com/groups/334039357095989.

199 **new experimental reversal surgeries:** Tommy Nai-Jen Chang et al., "Microsurgical Robotic Suturing of Sural Nerve Graft for Sympathetic Nerve Reconstruction: A Technical Feasibility Study," *Journal of Thoracic Disease* 12, no. 2 (February 2020): 97–104, https://doi.org/10.21037/jtd.2019.08.52.

200 **Botox injections in the armpit:** Nawrocki and Cha, "The Etiology, Diagnosis, and Management of Hyperhidrosis: Therapeutic Options."

200 **prescription drugs:** "Oral Medications—International Hyperhidrosis Society | Official Site," accessed April 25, 2019, https://sweathelp.org/hyperhidrosis -treatments/medications.html.

200 **microwave:** Nawrocki and Cha, "The Etiology, Diagnosis, and Management of Hyperhidrosis: Therapeutic Options." For an alarming first-person descrip-

tion, see: Scott Keneally, "Sweat and Tears," *New York Times*, December 7, 2011, Fashion, https://www.nytimes.com/2011/12/08/fashion/sweat-and-tears-first -person.html.

202 **iontophoresis:** Nawrocki and Cha, "The Etiology, Diagnosis, and Management of Hyperhidrosis: Therapeutic Options."

202 **secondary hyperhidrosis:** Hobart W. Walling, "Clinical Differentiation of Primary from Secondary Hyperhidrosis," *Journal of the American Academy of Dermatology* 64, no. 4 (2011): 690–95, https://doi.org/10.1016/j.jaad.2010 .03.013.

202 **curious case reported by two Milwaukee doctors:** Mark K. Chelmowski and George L. Morris III, "Cyclical Sweating Caused by Temporal Lobe Seizures," *Annals of Internal Medicine* 170, no. 11 (2019): 813–14, https://doi.org/10 .7326/L18-0425.

204 **it killed people in their prime:** J. F. C. Hecker, *The Epidemics of the Middle Ages*, 3rd ed., trans. B. G. Babington (London: Trübner, 1859), http:// wellcomelibrary.org/item/b2102070x.

204 **"For in the latter end of May . . . six hours":** Hecker, *Epidemics of the Middle Ages*.

205 **30% to 50% of people who contracted Sweate died:** Paul Heyman, Leopold Simons, and Christel Cochez, "Were the English Sweating Sickness and the Picardy Sweat Caused by Hantaviruses?," *Viruses* 6, no. 1 (2014): 151–71, https://doi.org/10.3390/v6010151.

205 **claiming only 1 in 100 escaped:** John Caius, *A Boke, or Counseill against the Disease Commonly Called the Sweate, or Sweatyng Sicknesse. Made by Ihon Caius Doctour in Phisicke. Very Necessary for Euerye Personne, and Muche Requisite to Be Had in the Handes of al Sortes, for Their Better Instruction, Preparacion and Defence, against the Soubdein Comyng, and Fearful Assaultying of the-Same Disease* (London: Richard Grafton, printer to the kynges maiestie, 1552), https://wellcomelibrary.org/item/ b21465290#?c=0&m=0&s=0&cv=0&z=-1.1388%2C-0.0829%2C3.2776%2C1 .6584. See also Hecker, *Epidemics of the Middle Ages*.

205 **"left London immediately . . . destiny at Tytynhangar":** Hecker, *Epidemics of the Middle Ages*.

205 **"Cause theim to lie on their right side, and bowe theim selves forward, call theim by their names, and beate theim with a rosemary braunche":** Caius, *A Boke*.

205 **"butter in a mornyng with sage":** Caius, *A Boke*.

205 **"figges before dinner":** Caius, *A Boke*.

205 **"Flesh meats highly . . . morning":** Hecker, *Epidemics of the Middle Ages*.

206 **"a total want of refinement in diet":** Hecker, *Epidemics of the Middle Ages*.

206 **"Queen Catherine had pot-herbs brought from Holland for the preparation of salads, as they were not procurable in England":** Hecker, *Epidemics of the Middle Ages*.

206 **Vesuvius:** Hecker in *Epidemics of the Middle Ages* claims that Mount Vesuvius erupted in 1506, which neither I nor my fact-checker can confirm. So perhaps it was just *active* or Hecker was misinformed about an eruption.

206 **wrath of God:** Hecker, *Epidemics of the Middle Ages*.

206 **influenza:** M. Taviner, G. Thwaites, and V. Gant, "The English Sweating Sickness, 1485–1551: A Viral Pulmonary Disease?," *Medical History* 42, no. 1 (1998): 96–98.

206 **typhus, plague, yellow fever, botulism, and ergotism:** Heyman, Simons, and Cochez, "English Sweating Sickness and the Picardy Sweat."

206 **hantavirus attack:** Heyman, Simons, and Cochez, "English Sweating Sickness and the Picardy Sweat."

206 **"The mystery around . . . the truth":** Heyman, Simons, and Cochez, "English Sweating Sickness and the Picardy Sweat."

207 **"Such is the disease . . . nearing extinction' ":** Henry Tidy, "Sweating Sickness and Picardy Sweat," *British Medical Journal* 2, no. 4410 (July 14, 1945): 63–64.

207 **cabaret career opportunity:** *Electric Man: France's Got Talent 2016—Week 5*, 2016, video, https://www.youtube.com/watch?v=QPDoqnPBVX4.

207 **Pajkic was born without sweat glands:** *Biba Struja (Battery Man)*, documentary directed by Dusan Cavic and Dusan Saponja (Ciklotron d.o.o., This and That Productions, 2012).

208 **"Every man is born with a purpose . . . cannot hurt me":** *Biba Struja (Battery Man)*.

208 **he cooks a sausage:** *Electric Man: France's Got Talent 2016—Week 5*.

208 **electrical engineer Mehdi Sadaghdar has shown:** Mehdi Sadaghdar, *Electrical Tricks of Biba Struja the Battery Man*, YouTube, ElectroBOOM, 2016, video, https://www.youtube.com/watch?v=Lh6Ob1HFC6k.

208 **"Both my kids . . . like ducks":** *Biba Struja (Battery Man)*.

209 **discharges electricity to his patients':** *Biba Struja (Battery Man)*.

209 **in utero between week 20 and week 30:** Holm Schneider et al., "Prenatal Correction of X-Linked Hypohidrotic Ectodermal Dysplasia," *New England Journal of Medicine* 378, no. 17 (2018): 1604–10, https://doi.org/10.1056/NEJMoa1714322.

209 **1 in 25,000 people are born with XLHED:** Antonio Regalado, "In a Medical First, Drugs Have Reversed an Inherited Disorder in the Womb," *MIT Technology Review*, April 25, 2018, https://www.technologyreview.com/s/611015/in-a-medical-first-drugs-have-reversed-an-inherited-disorder-in-the-womb/.

209   **inject the babies with functional versions:** Regalado, "In a Medical First."
209   **researchers in Erlangen, Germany, reported:** Schneider et al., "Prenatal Correction."
209   **A woman in her thirties:** Schneider et al., "Prenatal Correction."
210   **"We were hesitant . . . may bring":** Regalado, "In a Medical First."
211   **"Treating babies . . . pregnant woman":** Regalado, "In a Medical First."
211   **"If you wanted to make this for the patient community . . . three out of three":** Regalado, "In a Medical First."

11: SWEAT STAINS

I am so grateful to the conservation scientists, conservators, and curators who have spoken to me, including Renée Dancause, Michelle Hunter, Janet Wagner, Jonathan Walford, and Lucie Whitmore.

212   **aluminum components, such as the suit-wrist connectors:** Sarah Everts, "Saving Space Suits," *C&EN*, May 9, 2011.
213   **"It may look cool, but it's 35 years old, smells like a locker room and there's some discoloration on the inside":** Leah Crane, "Cosmic Couture: The Urgent Quest to Redesign the Spacesuit," *New Scientist*, January 3, 2018, https://www.newscientist.com/article/mg23731591-100-cosmic-couture-the-urgent-quest-to-redesign-the-spacesuit/.
213   **"The Pits of Despair?":** Anna Hodson, "The Pits of Despair? A Preliminary Study of the Occurrence and Deterioration of Rubber Dress Shields," in *The Future of the 20th Century: Collecting, Interpreting and Conserving Modern Materials: 2nd Annual Conference, 26–28 July 2005*, ed. Cordelia Rogerson and Parl Garside (London: Archetype, 2006).
213   **"belonging to one of the most privileged social classes of the European courts":** A. Hernanz, "Spectroscopy of Historic Textiles: A Unique 17th Century Bodice," in *Analytical Archaeometry: Selected Topics*, ed. G. M. Edwards and P. Vandenabeele (London: Royal Society of Chemistry, 2012).
213   **it's often acidic, with a pH as low as 4.5. As sweat decomposes, its pH rises past neutral 7 and into the alkaline, where it then dries:** Melanie Sanford and Margaret Ordonez, "The Identification and Removal of Deodorants, Antiperspirants, and Perspiration Stains from White Cotton Fabric," in *Strengthening the Bond: Science & Textiles*, ed. Virginia J. Whelan and Henry Francis du Pont (Philadelphia: North American Textile Conservation Conference, 2002), 119–31.
213   **"The longer dried perspiration stays on the fabric, the higher the degree of damage":** Sanford and Ordonez, "Identification and Removal."

214 **"There are also many examples where insects have preferentially eaten underarm and crotch areas"**: Jessie Firth, "Re: Media Request: Conservation of Sweat Stains on Textiles," email, February 2, 2012.

214 **modern antiperspirant formulations can combine with soap or detergent to form a discolored, brittle crust on fabric—especially cotton—that doesn't dissolve away in water**: Sanford and Ordonez, "Identification and Removal."

214 **The first time I visited the institute**: Sarah Everts, "Conserving Canada's Valuables," *Artful Science* (blog), May 23, 2011, https://cenblog.org/artful -science/2011/05/23/conserving-canada's-valuables/.

216 **chemical interactions of light and oxygen with the sweat's lactic acid**: Sanford and Ordonez, "Identification and Removal."

216 **World War II wedding dress**: *Fabric Preservation and the Application of Fake Sweat*, video, 2009, https://www.youtube.com/watch?v=7AJQKMYAltQ.

INDEX